Night Sky

MONTH-BY-MONTH

JANUARY–DECEMBER 2004

A FIREFLY BOOK

Published by Firefly Books Ltd. 2003

First printing

Publisher Cataloguing-in-Publication Data (U.S.)

Heudier, Jean-Louis.
    The night sky month-by-month : January - December 2004 / Jean-Louis Heudier ; with the collaboration of Charles Frankel – 1st ed.
[240] p. : photos. (chiefly col.), maps ; cm.
Includes bibliographical references.
Summary: Night-by-night reference to celestial phenomena in the Northern Hemisphere sky.
ISBN 1-55297-816-8 (pbk.)
1. Astronomy -- Observers' manuals. 2. Constellations -- Observers' manuals.
3. Stars -- Observers' manuals. I. Frankel, Charles. II. Title.
523.8/022/2 21   QB63.H47 2003

National Library of Canada Cataloguing in Publication Data

Heudier, J. L.
   The night sky month-by-month : January-December 2004 / Jean-Louis Heudier ; with the collaboration of Charles Frankel.
Includes bibliographical references.
ISBN 1-55297-816-8
   1. Astronomy--Observers' manuals. I. Frankel, Charles II. Title.
QB63.H49 2003   520'.22'3   C2003-902848-8

Published in the United States in 2003 by
Firefly Books (U.S.) Inc.
P.O. Box 1338, Ellicott Station
Buffalo, New York 14205

Published in Canada in 2003 by
Firefly Books Ltd.
3680 Victoria Park Avenue
Toronto, Ontario, M2H 3K1

Cover design: Tinge Design Studio

Printed in Canada by Friesens, Altona, Manitoba

Jean-Louis Heudier
with the collaboration of
Charles Frankel

THE Night Sky
MONTH-BY-MONTH

JANUARY–DECEMBER 2004

FIREFLY BOOKS

Foreword

A starry night is a wonderful sight. It is an open window onto thousands of stars of different shapes and colors, and onto the Universe in all its immensity. But on entering this Universe, you do not want to feel disorientated, lost and ignorant. This guide proposes to accompany you on an easy yet fascinating exploration of the sky. Month by month and day by day, it will point out what there is to observe as long as the sky is clear: the Moon and planets, showers of shooting stars, stars in their constellations, galaxies, etc.

From these thousands of objects that populate the sky, you will soon find your bearings and recognize familiar signs that will guide you on your voyage of discovery: the Pole Star, the Summer Triangle, particular constellations, and so on. Detailed photos, charts and notes are there to help you.

The panorama of the sky becomes fascinating if you are able to observe it with an astronomical instrument. Practical advice and explanations will tell you what you can expect from the different optical instruments, binoculars and telescopes designed for amateurs.

Finally, all these wonders will not fail to arouse many questions: how was our Earth formed, when will the Sun die, how far away is the nearest galaxy, what is the age of the Universe? There are articles for beginners in astronomy, and reports on current research will give you the facts and keep you up to date.

For the reader who already has some knowledge of the subject, this book carries all the information you need to indulge your passion: ephemerides, forecasts of meteor showers, eclipses and occultations.

From January through December 2004, the book accompanies you along the way, hoping that you will spend a year with your head up in the stars.

Light and shadow
The Hubble Space Telescope enables you to reach mysterious regions of the Universe. This is the Carina Nebula.

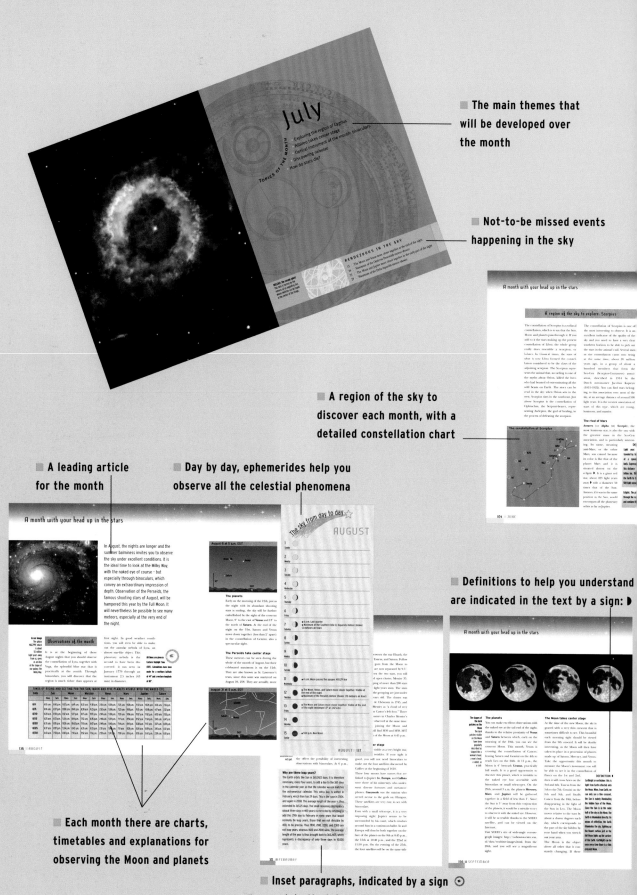

July

TOPICS OF THE MONTH
- Exploring the region of Cygnus
- Albireo takes center stage
- Optical instrument of the month: binoculars
- Discovering nebulae
- How do stars die?

The main themes that will be developed over the month

RENDEZVOUS IN THE SKY

Not-to-be missed events happening in the sky

A month with your head up in the stars

A region of the sky to explore: Scorpius

A region of the sky to discover each month, with a detailed constellation chart

A leading article for the month

Day by day, ephemerides help you observe all the celestial phenomena

A month with your head up in the stars

The sky from day to day

AUGUST

Definitions to help you understand are indicated in the text by a sign: ▶

A month with your head up in the stars

Each month there are charts, timetables and explanations for observing the Moon and planets

Inset paragraphs, indicated by a sign ⊙ complete the information

THE Night Sky MONTH-BY-MONTH

Instructions for use

■ **Each month, you get a general sky chart**

■ **Each month, there is practical advice to ensure your observations are successful**

■ **Each month, a topic introduces you to astronomy, followed by an article informing you about current research**

ALSO FOR YOU TO CONSULT: PRACTICAL INFORMATION (P. 226) AND GLOSSARY (P. 230).

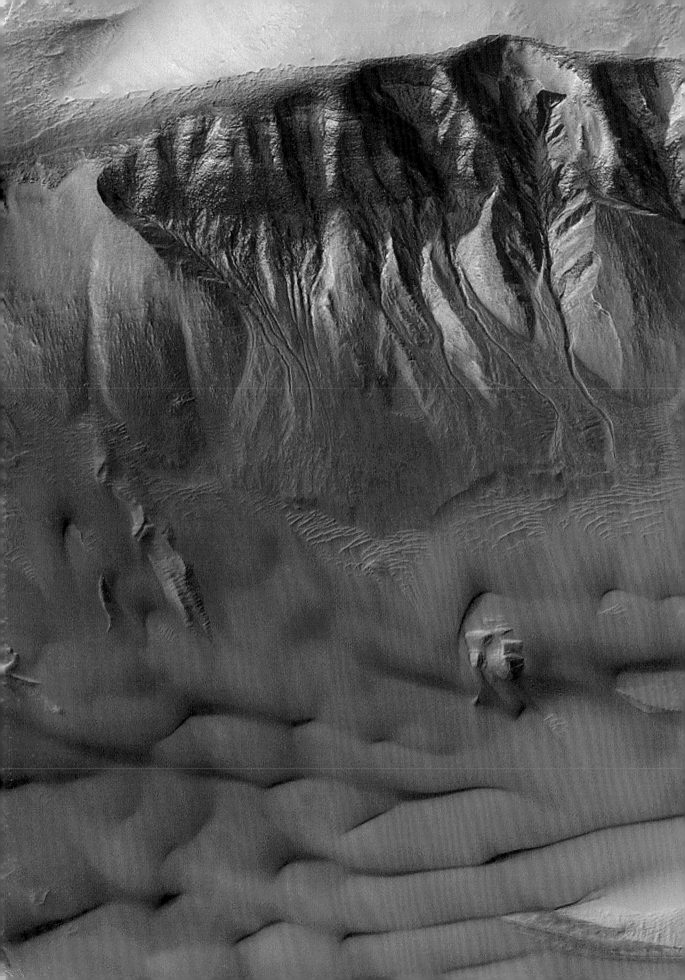

January

Canyons on Mars
Mars Global Surveyor has enabled us to discover landscapes related to the geological formations on our own planet.

A month with your head up in the stars

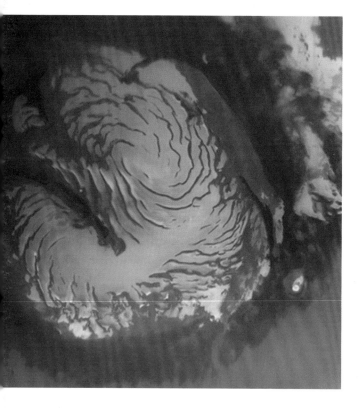

The planet Mars occupies center stage in January, since it is clearly visible on all evenings throughout the month. This red planet does not seem to follow the stars' slow daily shift in position. If you try to memorize how the sky looks each day, the planets' apparent movement in relation to the "fixed" stars will be revealed to you.

The pole of the planet Mars
The Mars Global Surveyor pictures show a seasonal development of frost cover, comparable to that found at the poles on Earth.

Observations of the month

Sirius is resplendent in the south: together with Betelgeuse and Capella, it offers a beautiful Winter Triangle for observers to admire. The days are beginning to get longer now, with the Sun slowly leaving the Tropic of Capricorn to move north again. Even though, at the beginning of the month, the Earth is as close as it possibly could be to the Sun, its passage through the perihelion is not accompanied by an increase in temperature. On the contrary, thermal inertia continues to bring low temperatures, despite the lengthening days and the Sun's relative proximity.

All times are given in Eastern Standard Time (EST). Calculations have been made for a latitude of 44° north and a longitude of 80° west.

TIMES OF RISING AND SETTING FOR THE SUN, MOON AND FIVE PLANETS VISIBLE WITH THE NAKED EYE														
Date	Sun		Moon		Mercury		Venus		Mars		Jupiter		Saturn	
	Rises	Sets	Rises	Sets	Rises	Sets	Rises	Sets	Rises	Sets	Rises	Sets	Rises	Sets
1/1	7:56 a.m.	4:50 p.m.	1:07 p.m.	2:17 a.m.	6:51 a.m.	4:06 p.m.	9:56 a.m.	7:32 p.m.	11:54 a.m.	12:30 a.m.	10:35 p.m.	11:22 a.m.	4:40 p.m.	7:57 a.m.
1/5	7:56 a.m.	4:53 p.m.	3:06 p.m.	6:30 a.m.	6:27 a.m.	3:44 p.m.	9:54 a.m.	7:43 p.m.	11:43 a.m.	12:27 a.m.	10:20 p.m.	11:06 a.m.	4:23 p.m.	7:40 a.m.
1/10	7:55 a.m.	4:59 p.m.	8:09 p.m.	10:07 a.m.	6:14 a.m.	3:27 p.m.	9:49 a.m.	7:56 p.m.	11:29 a.m.	12:24 a.m.	10:00 p.m.	10:47 a.m.	4:01 p.m.	7:19 a.m.
1/15	7:53 a.m.	5:05 p.m.	12:52 a.m.	11:58 p.m.	6:13 a.m.	3:19 p.m.	9:44 a.m.	8:09 p.m.	11:16 a.m.	12:21 a.m.	9:39 p.m.	10:27 a.m.	3:40 p.m.	6:58 a.m.
1/20	7:50 a.m.	5:11 p.m.	7:13 a.m.	3:34 p.m.	6:19 a.m.	3:19 p.m.	9:37 a.m.	8:21 p.m.	11:03 a.m.	12:17 a.m.	9:18 p.m.	10:07 a.m.	3:18 p.m.	6:36 a.m.
1/25	7:46 a.m.	5:18 p.m.	10:12 a.m.	9:51 p.m.	6:27 a.m.	3:25 p.m.	9:30 a.m.	8:34 p.m.	10:50 a.m.	12:14 a.m.	8:57 p.m.	9:48 a.m.	2:57 p.m.	6:15 a.m.
1/30	7:42 a.m.	5:25 p.m.	11:56 a.m.	2:13 a.m.	6:36 a.m.	3:35 p.m.	9:22 a.m.	8:46 p.m.	10:37 a.m.	12:12 a.m.	8:36 p.m.	9:26 a.m.	2:36 p.m.	5:55 a.m.

January 9 at 6 a.m. EST

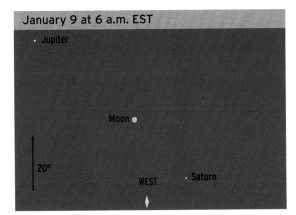

- Jupiter
- Moon
- 20°
- WEST · Saturn

The planets

The beginning of the month sees the Moon moving away from **Mars** and toward **Saturn**. **Venus**, on the Western horizon, seems to become higher from day to day and get closer to Mars. The ecliptic can be seen very clearly in the sky. On the 14th, Venus, in the south, is crossing the planet **Uranus**. On both the 14th and 15th, it is quite easy to find Uranus with binoculars: it is just 1° above Venus.

On the 24th, a lovely crescent Moon of a light silvery-gray color passes 3° from Venus. From the 24th to the 28th, the Moon travels along the arc connecting Venus to Mars, creating beautiful configurations that can be seen early in the evening. You should try to take some very simple photographs of them to

January 27 at 7 p.m. EST

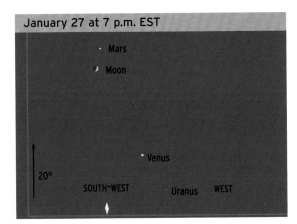

- · Mars
- ☽ Moon
- · Venus
- 20°
- SOUTH-WEST   Uranus   WEST

1 Thursday	◖	
2 Friday	◖	
3 Saturday	◖	■ 3 p.m. Moon passes the apogee: 405,707 km
4 Sunday	◖	■ 4 p.m. Earth passes the perihelion: 0.98325 AU (147,092,106 km) ■ Maximum of the Quadrantid meteor shower (50 meteors an hour)
5 Monday	◗	■ 7:26 p.m. Minimum of Algol
6 Tuesday	◗	■ The Moon and Saturn move closer together. Can be seen all night long (minimum 4.5° at 7 p.m. on the 5th)
7 Wednesday	◗	■ 10:43 a.m. Full Moon
8 Thursday	◗	■ 4:15 p.m. Minimum of Algol
9 Friday	◗	
10 Saturday	◗	
11 Sunday	◗	■ 1:05 p.m. Minimum of Algol
12 Monday	◗	■ The Moon and Jupiter move closer together. Can be seen all night long (minimum 3.4° at 6 a.m.)
13 Tuesday	◗	
14 Wednesday	◗	■ 11:45 p.m. Last quarter of the Moon
15 Thursday	◗	

The planet
Mars
Will human
beings soon
reach Mars?

register the change in the relative positions of the moving stars. On the 14th, Mars is 0.5°, that is the diameter of the Moon, to the south of Epsilon Pisces, a star that is visible with the naked eye (magnitude 4.3).

Observe the way they are moving closer together, both through binoculars and with your naked eye. On the 29th, the planet crosses the line linking Omicron (2°) and Eta (4.5°) in Pisces. You will thus be able, in the space of two weeks, to

measure the movement of the planet Mars in relation to the stars, and in particular to the constellation of Pisces, one of the least conspicuous in the zodiac.

Mars center stage

Mars appears to be moving backward among the constellations, and coming closer to Orion – the god of war is approaching the great hunter in the sky. It was while searching for the secret of

Mars in figures

Distance to Sun	1.523 AU, 227,936,640 km
Distance from perihelion	1.38 AU, 206,600,000 km
Distance from aphelion	1.67 AU, 249,200,000 km
Eccentricity of orbit	0.093
Inclination of orbit	1.85°
Equatorial diameter	6,787 km (0.532 Earth)
Mass	6.41×10^{23} kg (0.107 Earth)
Revolution period	687 days
Orbital velocity	24.13 km/s
Rotation period	24 hours 37 minutes
Escape velocity	5.02 km/s

this trajectory that Kepler showed the planet was not moving in a circular orbit but in an ellipse, of which the Sun occupied not the center, but a focus.

Observe Mars through binoculars and note as precisely as possible the evolution of its movement from day to day among the constellations. With a small to medium telescope, you will be able to see the planet's features, in particular the polar caps. In 1784, William Herschel was the first to advance the theory that, since Mars and the Earth revealed similar characteristics, the white areas visible at the poles might be frozen areas identical to the poles on Earth.

Shooting stars

A feature of the start of the year is the **Quadrantids** meteor shower, with a maximum on the 4th. The constellation of Quadrans Muralis, which gave its name to this meteor shower, no longer exists. It was between Hercules' right foot, the Herdsman's left hand and Draco. It is thus from the constellation of Draco that these shooting stars appear to come.

16 Friday

17 Saturday — 4 a.m. Greatest western elongation of Mercury (23.9°)

18 Sunday

19 Monday — 2 p.m. Moon passes perigee: 362,768 km

20 Tuesday — Mercury and the Moon move closer together. Visible in the morning (minimum 4.7° at 10 p.m. on the 19th)

21 Wednesday — 4:08 p.m. New Moon

22 Thursday

23 Friday

24 Saturday — The Moon and Venus move closer together. Can be seen at the beginning of the night (minimum 3.6° at 11 a.m.)

25 Sunday

26 Monday

27 Tuesday — The Moon and Mars move closer together. Can be seen at the beginning of the night (minimum 2.6° at 10 p.m.)

28 Wednesday

29 Thursday — 1:30 a.m. First quarter of the Moon

30 Friday

31 Saturday — 9 a.m. Moon passes the apogee: 404,807 km

A month with your head up in the stars

Exploring a region of the sky: Ursa Major

The Earth's axis of rotation is nowadays pointed toward Polaris. The sky seems to revolve around this star, the only one to appear motionless. However, we now know that this direction of the Earth wavers slightly over time: in a few thousand years, the star Vega (Alpha Lyrae) will indicate the north.

Seven stars stand out in the great merry-go-round that makes up the night sky. They are always visible in the northern hemisphere, at any time of night and at all times of the year. Thanks to them, we can easily spot the only star that seems fixed in the sky, Polaris, which always indicates a northerly direction and stands as a symbol for the order that lies beneath the apparent disorder in the myriad of stars. In former days, the north was naturally associated with the seven bright stars in Ursa Major. To Greek and Babylonian eyes they formed the shape of a bear, while for the Romans they seemed to represent oxen at work, and for the northern Europeans a great chariot or saucepan. In North America, it is called the Big Dipper, in Britain, the Plow.

The navigators of antiquity were able to orient themselves thanks to Ursa Major, as well as its neighbor, Ursa Minor. Such is its importance that the names given to it have remained engraved in our cultures and continue to be used to designate the north. The Greek word for "bear" was **arctos** and we have kept the term "arctic"; the Romans called the seven laboring oxen **septem triones** and the now somewhat archaic word "septentrion" used to be used to mean the north.

Numerous legends are attached to this important constellation, the most renowned of which is that of Callisto, the beautiful nymph who was an attendant of Artemis. Zeus, using one of his well-known tricks, managed to make her pregnant and she gave birth to a child by the name of Arcas. Hera (the wife of Zeus) turned the beautiful Callisto into a bear, leaving young Arcas to be taken care of by Artemis' companions. When he had grown up and become a hunter, Arcas was one day at the point of killing his mother, whom he did not recognize. This would have achieved the revenge desired by Hera, but Zeus halted the murder by turning Arcas into a bear as well, and placing the two animals among the stars. This is why the great bear, Callisto, is always accompanied by her son, the little bear. However, Hera was still angry and, with the help of Poseidon, she moved the two bears toward the north, to a region of the sky in which they were condemned to be forever visible, which meant never being able to rest.

The second star of the saucepan handle is a double one, Alcor and Mizar. This double star plays various roles in different cultures, representing either the coachman of the chariot, or the hunter making off with the pot meant for cooking the bear in. This star is a good test of observation because even with the naked eye two stars are visible. At different periods in history, the ability to discern these two stars served as a recruitment test for archers.

Dubhe (or **Alpha (α) Ursae Majoris**), derives its name from the Arabic *thahr al dubbal akbar*, "the rear of the great bear."

THE SKY IN JANUARY

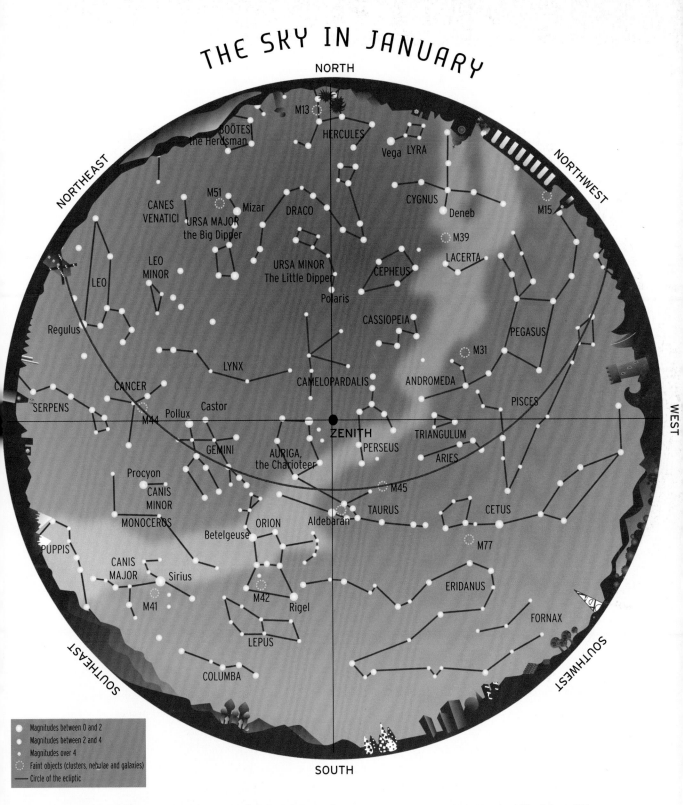

NORTH

NORTHEAST

NORTHWEST

M13

BOÖTES
the Herdsman

HERCULES

Vega   LYRA

CANES
VENATICI

M51

Mizar

DRACO

CYGNUS

Deneb

M15

URSA MAJOR
the Big Dipper

M39

LACERTA

LEO
MINOR

URSA MINOR
The Little Dipper

CEPHEUS

PEGASUS

LEO

Polaris

CASSIOPEIA

Regulus

M31

ANDROMEDA

LYNX

CAMELOPARDALIS

PISCES

SERPENS

CANCER

Castor

ZENITH

TRIANGULUM

M44   Pollux

PERSEUS

ARIES

GEMINI

AURIGA,
the Charioteer

Procyon

M45

CANIS
MINOR

TAURUS

CETUS

MONOCEROS

ORION

Aldebaran

PUPPIS

Betelgeuse

M77

CANIS
MAJOR   Sirius

ERIDANUS

M41

M42

Rigel

FORNAX

LEPUS

COLUMBA

WEST

SOUTHEAST

SOUTHWEST

SOUTH

- ○ Magnitudes between 0 and 2
- • Magnitudes between 2 and 4
- · Magnitudes over 4
- ⊙ Faint objects (clusters, nebulae and galaxies)
- — Circle of the ecliptic

■ How to use this chart
Hold the chart above your head, matching the word SOUTH that appears at the edge of it with the geographical south of the place you are observing from. Use a compass to help you do this.

■ This chart shows the sky that is visible at a latitude of 45°
If you are further north or further south, Polaris will be higher or lower in the sky.

■ Chart of the sky visible at 11 p.m. EST at the beginning of the month; at 10 p.m. EST in the middle of the month; at 9 p.m. EST at the end of the month.

With Beta (β), it forms the end of the saucepan, together with the two "pointers" that allow the Pole Star to be located. This is the first bright star to be found by following a line from the two pointers (α and β), at a distance equal to five times their distance from one another. You can also find the Pole Star in the following way: the angular distance between it and Dubhe is 28°, or the length of your arm with the fingers of your hand spread out. If you observe Dubhe with the aid of binoculars or a small telescope, you will soon discover that it has a companion,

with a magnitude of 7 to 6.3. This star appears blue, whereas in fact it is yellow. This is due to a contrast effect produced by the principal star, which has a very pronounced golden yellow color.

Mizar and **Alcor** (or **Zeta** (ζ) **Ursae Majoris**) form one of the most renowned groups in the sky. Mizar is reminiscent of a horse, with the second star the rider. They are not really a pair, but two stars set at very different distances (60 light years away for Mizar, 90 for Alcor) and Alcor is not as bright as Mizar. They are spectacular to observe, first through

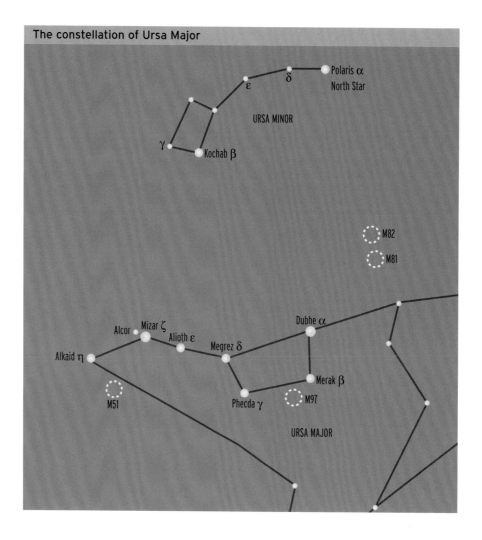

The constellation of Ursa Major

Doppler-Fizeau effect: this is the distortion of light caused by movement of its source. When the source is approaching the observer, the color of the light shifts towards the blue, when the source is receding, the color of the light shifts towards the red. When the light is decomposed into a spectrum, all the lines making up that spectrum shift towards blue when the source is approaching, and towards red when it is receding.

binoculars, then with the aid of a more powerful instrument.

Mizar has a very close companion: Giovanni Riccioli identified it in 1650, and was thus the first to discover a pair of stars linked by gravitation. This pair was the first to be photographed, by George Bond, in 1867. It was also by observing this pair that Edward Pickering, in 1889, was able to reveal for the first time the existence of what he called "double spectroscopic stars." When two stars revolve permanently around one another, one of them approaches the Earth and the other moves away from it; because of the Doppler-Fizeau effect ▶, the first presents a spectrum verging toward blue and the other toward red. When these stars are very near to one another, the spectrum of the whole results from the blending of the two individual spectra, which explains the term "double spectroscopic." Thus the period of the pair can be assessed: you simply need to measure the time taken by the spectrum to unfold, and then to return to normal.

M81 and M82
One of the galaxies, on the right, is viewed in section, while the other is observed.

Three galaxies to observe

M81 and **M82**, two beautiful galaxies discovered by Johan Bode in 1774, are situated 12 million light years from the Sun. They are easy to find with a small optical instrument, especially M82, which is brighter than its neighbor. A fairly simple method consists of extending the diagonal of the saucepan toward the north, that is from Gamma (γ) to Alpha (α), by a length equal to the distance separating these two stars. M81 and M82 are just to the north of the point reached. It is rare to be able to observe two galaxies close together in the same part of the sky.

M51, the renowned Whirlpool galaxy, is at the edge of Ursa Major. It is one of the most beautiful spirals that can be observed with the aid of a just small instrument. Discovered by Charles Messier on October 13, 1773, it is 37 million light years away; its companion NGC5195 (see picture) was located by Méchain in 1781. The spiral structure of M51 was demonstrated by Lord Ross in 1845. It is easy to find from Alkaid, the star at the end of the Big Dipper's handle. You simply need to come back 2° toward the body of the bear (toward the west) and point at a small star, 24 Uma: M51 is 2° to the south of this star.

M51
This beautiful spiral galaxy, known as the Whirlpool, has a close companion with which it interacts.

Practical astronomy

How to photograph the sky without a telescope

It is possible for you to take excellent astronomical photographs without any kind of telescope. You can do a lot with a simple camera, assuming that you observe a few important rules.

With limited exposure time

You can photograph the sky that you see without difficulty with a few seconds' exposure. However, the action of the Earth's rotation causes all the stars to appear to move and, if you don't want to see trails left by objects in their apparent movement around the Pole, you must take a few precautions. Taking into account the focal length of the camera and the direction of the sky it is aimed at, you can adjust the exposure time so that the stars appear clearly defined and the picture gives the impression of an instant snapshot. An element of surprise generally arises from the fact that the number of stars appearing on the negative is much greater than the number it is possible to distinguish with the naked eye. To produce this kind of picture, you need to have a camera fitted with T or B exposure. T exposure, which has become rare on modern cameras, operates in two stages: when you press the shutter release button for the first time, the lens opens; when you press it for the second time, it closes and the exposure ends. B exposure allows you to keep the shutter open while pressing down the shutter release button. In this case, it is preferable to use a shutter release cable fitted with a locking system so that the camera does not move during exposure. Keep experimenting in order to find out the best conditions for taking photographs. As a guideline, to photograph the area around the celestial equator with a lens of 50 mm focal length, 15 seconds' exposure gives good results. For greater latitudes, the exposure time needs to be increased, until it reaches almost one minute near the Pole. This type of snapshot enables you to reproduce the constellations with ease, to photograph the planets in their zodiacal constellations and even, with luck, to capture shooting stars.

With extended exposure time

If you leave the shutter open for longer, you will get star trails which show the

The dance of the stars
An exposure of almost six hours allows the rotation of the stars around the North Pole to be shown.

Earth's rotation. This kind of photograph is easy to obtain and is more spectacular the longer the exposure. Sometimes, in addition to the concentric circles from star trails, you even get shooting stars or bolides recorded during the exposure – a heavenly surprise! If your camera is safely set up on a tripod, you can leave the shutter open for several hours, as long as it is not too modern a camera and there is no way of deactivating the automatic exposure control system. In any case, such systems consume a great deal of power and very quickly empty the camera's batteries. To produce the required images, it is therefore better to have an old-fashioned, completely mechanical camera, perhaps equipped with T exposure. Unfortunately, such "out-of-date" equipment is becoming increasingly difficult to find.

A Moon halo
This halo of 22° around the Moon stems from the diffusion of light by ice crystals present in the upper atmosphere.

A polar dawn

Spectacular phenomena

Aurora borealis (The northern lights)

Polar lights are not only to be found in northerly regions. Magnificent auroras have been observed much further south, so get out your cameras! A few seconds' exposure is enough to capture these lights. It is, however, useful to remember that this phenomenon is caused by atoms of atmospheric gas that are stimulated by electrical particles from the Sun. The colors derive from very precise wavelengths, and photographic emulsions do not necessarily respond to these in the same way as our eyes do. So the photograph may only bear a faint resemblance to the phenomenon observed.

Eclipses

For eclipses, of the Sun or of the Moon, there are two possible types of picture. If the Sun or Moon is itself the focus of interest, you need to use a telephoto lens or an astronomical lens. If you want to convey the very special ambience produced by these phenomena, then photograph the Sun or Moon against the landscape. (Take care never to look directly at the Sun - especially not through a telescope or camera lens - even when it is partially obscured by the Moon. A reflector telescope is ideal for observation of the Sun and taking photographs of it.) The results can be very interesting, but we should not forget that the image of the Sun or Moon only measures, through the camera's focusing, one hundredth of the focal length: a 50 mm lens gives a picture of the Sun or the Moon which is 0.5 mm in diameter. Photographs showing the moon in its different positions during the course of the night are great fun to take, but they require practice and equipment capable of superimposing on each of the exposures, neither of which is very common.

The four telluric planets Starting from the Sun, the nearest planets are, in order: Mercury, Venus, the Earth and Mars.

Discovering the telluric planets

In our solar system, the first four planets (Mercury, Venus, Earth and Mars) are rocky ones. They are rich in minerals but poor in gases, and they owe this composition to the fact that they were born near to the Sun, where there was an abundance of silicates and metal oxides, whereas volatile elements like hydrogen and helium had been driven out by thermal agitation. Small specks, measuring only a few millimeters, then stuck together in blocks of several kilometers. The blocks collided with one another and, over the course of a few million years, produced these four telluric planets.

Mercury

Mercury, the planet closest to the Sun (at an average distance of 58 million kilometers), is mainly composed of refractory elements such as iron and nickel, assem-

bled in a dense central core (70% of the planet's mass), and covered with a mineral coating and a fine lava crust on the surface. It is the smallest of the telluric planets (apart from Pluto), with an equatorial diameter of 4,878 kilometers. Mercury is hardly any bigger than our Moon, which it resembles on account of its uneven surface. Its small size meant that it quickly lost its internal heat. Apart from some volcanic activity very early in its history, geological activity on Mercury has ceased. Another consequence of its small size is that this planet has a very weak force of gravity (0.38 G, G being the Earth's gravity). Along with the very warm environment that accompanied its formation, this low gravity prevented gases accumulating around Mercury, so that the planet has no atmosphere. Contrasts in temperature are therefore very marked: 620 K the average during the day (the maximum measured being 700 K) and 100 K at night. The days and nights are long: Mercury rotates completely only every 58.5 days.

Because it is very near the Sun, it is difficult to observe Mercury through a telescope from Earth. We had to wait for the Mariner 10 probe to fly over it, in 1974–75, for us to discover a good half of its surface in detail. This surface is riddled with impact craters, of which the widest, known as the Caloris Basin, forms a scar 1,300 kilometers in diameter.

Venus

Venus is the second telluric planet in order of distance from the Sun. With an almost circular orbit (which takes 225 days) 108 million kilometers from the central star, Venus is Earth's closest neighbor (less than 40 million kilometers away). Venus therefore appears especially bright (it is also known as the "Evening Star"), particularly since it is completely covered with clouds reflecting the Sun's light. This veil of cloud, on the other hand, hides its surface, and only by means of probes equipped with radar (Pioneer in 1978, Venera 15 and 16 in 1983, and especially Magellan in 1990–94) have we been able to discover the turbulent planet beneath, furrowed with fractures and scattered with volcanoes.

The atmosphere of Venus is fascinating. The ground temperature reaches 750 K, both night and day, a furnace resulting from a strong greenhouse effect. Venus' atmosphere is in fact made up of a thick layer of carbon dioxide (95%) and nitrogen (3%), with pressure approaching 100 bars on the ground – in other words, the same pressure that is found in Earth's oceans at depths of 1,000 meters. This inferno was investigated by about 10 Soviet probes between 1969 and 1984. Even though the first ones were literally carbonized and crushed by the pressure before touching the ground, Venera 7, 8, 9 and 10 were able to send back photographs revealing fine strata of volcanic rocks resembling basalt.

Mercury
This picture of the surface of Mercury's southern hemisphere was produced from images obtained by the Mariner 10 probe.

Venus
Below is a synthesis of radar observations carried out on Venus.

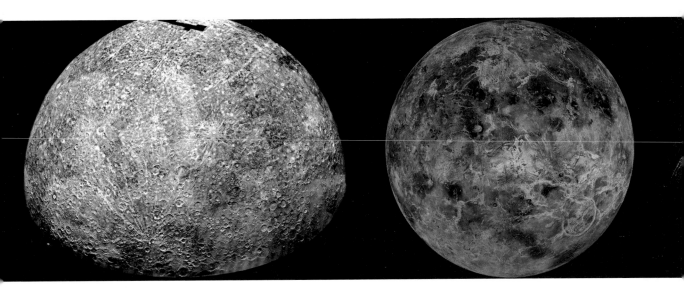

Venus is only slightly smaller than the Earth and, because of their comparable size, geologists were expecting to discover volcanic and tectonic activity on Venus that was similar to that on Earth. In fact, the radar pictures revealed a planet with an unfamiliar appearance, heavily fractured and with a superabundance of volcanoes (1,700 of them exceeding 20 kilometers in diameter), but no mountain chains or rifts or other alignments typical of plate tectonics on Earth.

The Earth

The Earth, situated 150 million kilometers from the Sun, is the largest of the telluric planets (see p. 182). It is also the only planet to possess an enormous satellite, the Moon (see p. 164). Mercury and Venus are without one and Mars only holds two small asteroids in its orbit: Phobos and Deimos.

Mars

Mars, the "red planet," is 220 million kilometers (on average) away from the Sun. Although it is half the size of the Earth in diameter, it offers an astonishing variety of terrains, revealed by space probes: former plateaus riddled with impact craters, and ravines in places where streams and rivers have dried up;

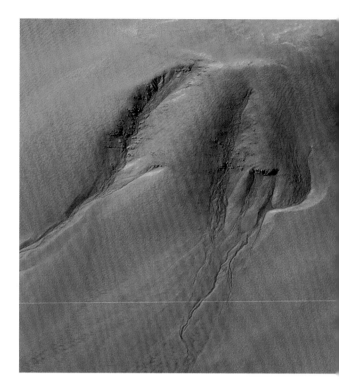

On Mars
This picture of ravines, taken by Mars Global Surveyor, reveals evidence of erosion similar to that caused by liquid water.

giant volcanoes (Olympus Mons measures 600 kilometers in diameter and is 27 kilometers high); an immense rift bursting through the crust at the level of the equator (the Valles Marineris, 4,000 kilometers long and 8 kilometers deep in places); flood channels which must have carried water flows 1,000 times greater than that of the Amazon; basins filled with sediment; glistening polar caps, combining water ice, mineral dust, carbon dioxide, and so on.

The atmosphere on Mars is thin: 6 to 7 hectopascals (millibars); in other words,

The telluric planets in figures				
	Mercury	Venus	Earth	Mars
Average distance from the Sun (millions km)	58	108	149	228
Period of revolution (days)	88	225	365	687
Equatorial diameter (km)	4,878	12,104	12,756	6,787
Rotation period	59 days	243 days	23 h 56 min	24 h 37 min
Density	5.41	5.25	5.52	3.93
Gravity (Earth = 1)	0.38	0.89	1.00	0.38

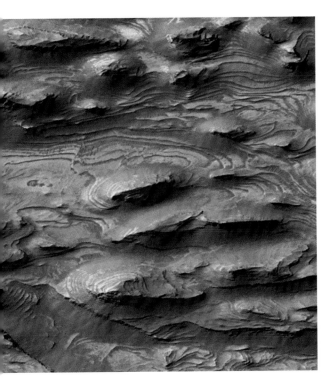

Water in liquid form cannot exist at present on the surface of Mars: the pressure is too weak and the temperature too low (225 K or –50°C on average). But there is no reason not to think that it may exist, compressed deep within rocks. It might periodically pour out onto the surface, before evaporating. Researchers have even put forward the theory that oceans once existed, at a time when the atmosphere was warmer and thicker.

The possibility that there was once, or is now, a primitive form of life is not excluded either. It is for all these reasons that, over the years to come, exploring the planet Mars will be a priority for the space agencies, and numerous missions have already been planned.

On Mars
The structure resembles the layers of sedimentation found on Earth.

the pressure encountered on Earth at an altitude of 35 kilometers. It is mainly composed of carbon dioxide (95%) and nitrogen (3%), with a few traces of water vapor.

Life on Mars?

The theory of intelligent life on Mars was put forward by the American Percival Lowell in 1894. Through his 18-inch (46 cm) aperture telescope, the astronomer thought he could see artificial channels on the surface of the red planet. Changes of color through the seasons were even attributed to a form of vegetation. In fact, space probes have revealed a sterile-looking, inorganic planet. The Viking probes (1976) were equipped with four devices for detecting life and organic molecules. Three gave negative results, but the fourth produced indicators suggesting the existence of microbes, detected through their respiration. A meteorite (ALH 84 001) from Mars, which fell into the Antarctic, also delivered up indicators of life, this time in fossil form: small spheres and short sticks, measuring less than 1 micron, resembled petrified bacteria. But they could also have been contamination by bacteria from Earth, which had filtered into the rock after its arrival. In order to resolve this debate, probes will need to bring back samples of rocks taken directly from Mars.

The sky tomorrow

Mission to Mars

Mars is the most popular planet for scientists and the general public. Despite recent pictures revealing an apparently sterile planet, the hope remains that Mars once sheltered a primitive form of life, and even that this still persists, hidden in ground water. Between now and 2015, therefore, the space agencies intend to set up a program of systematic exploration of the red planet by means of robots, collecting soil samples for the return to Earth. Beyond 2015, there is even talk of manned space flights.

This infatuation goes back to the early days of space conquest. Mars has been the target for about 30 space probes since 1961. Only nine missions, up to 2003, have been successful, all of them American probes. The first flight over Mars was accomplished by Mariner 4, on July 15, 1965: 22 black and white pictures, of mediocre quality, were transmitted. A leap forward was achieved by Mariner 9, which put itself into Martian orbit in November 1971: working from some 7,000 pictures, researchers discovered a new planet harboring giant volcanoes, rifts and dried-up valleys. In 1976, Viking 1 and 2 landed in the plains of the northern hemisphere, providing the first views of Martian soil.

Abandoned for about 15 years, the red planet once again became the target for American probes in the 1990s. Pathfinder became the third to make a gentle landing on Mars on July 4, 1997. Its little mobile robot, Sojourner, embarked on a walk of about 100 meters around the landing platform, carrying out a dozen chemical analyses of the rocks and soil along the way. Mars Global Surveyor went into orbit around Mars in the same year and managed to provide

Mars Express
This European probe, seen here in model form, was launched in June 2003.

On Mars
This picture, reminiscent of landscapes on Earth, was taken by Mars Global Surveyor.

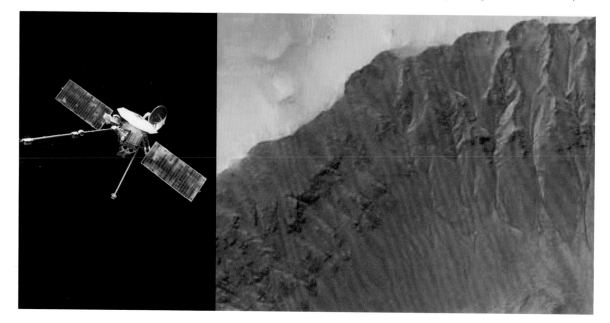

more than 100,000 pictures, as well as drawing up a topographical relief map by means of a laser altimeter.

Bringing back rock specimens

After the failures of Climate Observer and Polar Lander in 1999, Mars Odyssey met with success in 2001. Having entered orbit around Mars on October 23, it is in the process of drawing up a mineral chart of the planet, thanks to its infrared spectrometer. On the basis of this information, sites will be chosen for future landing missions, planned for 2003–2004. Two new American robots will examine the Martian soil, photographing and analyzing the rocks they come across. As for Europe, it will launch Mars Express in 2003, hoping to reach the planet at the end of the year. The probe will include a compact landing craft (less than 100 kilograms), which will analyze the soil looking for signs of life, and an orbiter equipped with radar to probe the areas flown over and detect any ground water that may exist.

France and the United States are jointly preparing a mission for the period 2007–2014 to bring back samples. American rovers will land on Mars to gather fragments of rock, which will be put into orbit by rocket, then picked up by a French computer-guided spacecraft responsible for returning them to Earth in 2014.

Human beings on Mars

NASA has drawn up plans for the future landing of people on Mars. Astronauts will have to undertake a dangerous mission lasting two and a half years: six months for the outward journey, six months for the return flight and one and a half years spent on Mars.

Unmanned modules will first of all land on Mars to prepare for the mission, including the re-ascent module to be used by the astronauts. The crew will only take off from Earth once these modules are in place. They will land near to the automatic modules and will carry out an in-depth exploration of the site, with the aid of a pressurized laboratory and vehicles for long forays.

At the end of their stay, the astronauts will use the re-ascent module, loaded with propellant fuel produced on the spot (methane and oxygen) and will rejoin the Martian orbit where the return spaceship will be waiting to take them back to Earth. If the political and economic circumstances are favorable, this odyssey could take place by around the year 2020.

Pathfinder and the Sojourner robot on Martian soil

February

Jupiter
Jupiter's atmosphere is turbulent, as is evident from the gaseous whirlwinds present throughout the equatorial region

RENDEZVOUS IN THE SKY

A month with your head up in the stars

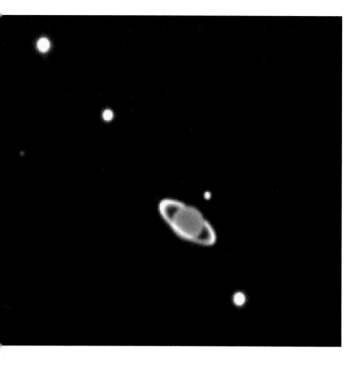

This month of February offers us the spectacle of Venus approaching Mars, which is itself moving closer to Saturn. At the end of the month, the trio is joined by the crescent-shaped Moon, at the moment when Jupiter is becoming clearly visible on the eastern horizon. In addition to this ballet of the planets, the zodiacal light is worth observing - if conditions are good.

Uranus
This picture, which looks like an amateur photo of Saturn, is in fact a view of the distant Uranus, surrounded by its rings and five of its satellites.

Observations of the month

The evening sky is dominated by the Winter Triangle, which is made up of Sirius, Procyon and Betelgeuse. There is a magnificent array of very luminous stars: Rigel and Aldebaran are in the southwest, and Castor and Pollux are reaching the zenith, while Arcturus is emerging on the eastern horizon.

The planets

Mars is emerging from the constellation of Pisces and moving toward Aries, while **Venus** is approaching Pisces. On the evening of the 27th, **Jupiter** is in the immediate vicinity of a

All times are given in Eastern Standard Time (EST). Calculations have been made for a latitude of 44° north and a longitude of 80° west.

TIMES OF RISING AND SETTING FOR THE SUN, MOON AND FIVE PLANETS VISIBLE WITH THE NAKED EYE														
Date	Sun		Moon		Mercury		Venus		Mars		Jupiter		Saturn	
	Rises	Sets	Rises	Sets	Rises	Sets	Rises	Sets	Rises	Sets	Rises	Sets	Rises	Sets
2/1	7:39 a.m.	5:28 p.m.	1:01 p.m.	4:19 a.m.	6:40 a.m.	3:41 p.m.	9:19 a.m.	8:54 p.m.	10:32 a.m.	12:11 a.m.	8:23 p.m.	9:19 a.m.	2:32 p.m.	5:47 a.m.
2/5	7:35 a.m.	5:34 p.m.	3:40 p.m.	7:37 a.m.	6:46 a.m.	3:54 p.m.	9:12 a.m.	9:04 p.m.	10:22 a.m.	12:09 a.m.	8:05 p.m.	9:03 a.m.	2:15 p.m.	5:30 a.m.
2/10	7:28 a.m.	5:41 p.m.	10:43 p.m.	9:40 a.m.	6:53 a.m.	4:12 p.m.	9:04 a.m.	9:15 p.m.	10:10 a.m.	12:06 a.m.	7:43 p.m.	8:42 a.m.	1:54 p.m.	5:10 a.m.
2/15	7:21 a.m.	5:48 p.m.	3:50 a.m.	12:18 p.m.	6:58 a.m.	4:34 p.m.	8:55 a.m.	9:27 p.m.	9:58 a.m.	12:04 a.m.	7:20 p.m.	8:21 a.m.	1:33 p.m.	4:49 a.m.
2/20	7:14 a.m.	5:54 p.m.	7:49 a.m.	6:18 p.m.	7:01 a.m.	4:58 p.m.	8:46 a.m.	9:38 p.m.	9:46 a.m.	12:01 a.m.	6:57 p.m.	8:00 a.m.	1:13 p.m.	4:29 a.m.
2/25	7:06 a.m.	6:01 p.m.	9:33 a.m.	11:58 p.m.	7:02 a.m.	5:25 p.m.	8:37 a.m.	9:49 p.m.	9:34 a.m.	11:58 p.m.	6:39 p.m.	7:40 a.m.	12:52 p.m.	4:09 a.m.
2/29	6:59 a.m.	6:06 p.m.	11:37 p.m.	4:04 a.m.	7:02 a.m.	5:48 p.m.	8:30 a.m.	9:58 p.m.	9:25 a.m.	11:56 p.m.	6:20 p.m.	7:23 a.m.	12:32 p.m.	3:51 a.m.

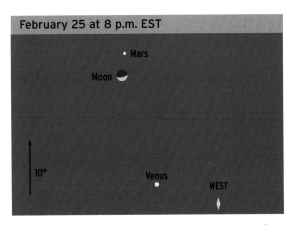

February 25 at 8 p.m. EST

Mars

Moon

10°

Venus

WEST

small star (magnitude 4.6) in the constellation of Leo. The apparent distance separating the two stars is only a few minutes of arc. Through binoculars, you will discover the richness of this field and the swarm of small stars in front of which Jupiter and its satellites are moving. On the evening of the 27th, at about 8 p.m., the four Galilean satellites are visible around the planet. This is a good opportunity to identify them: in order of distance from Jupiter, you will be able to distinguish Io, Europa, Ganymede and Callisto. This evening of the 27th also reveals a near-perfect alignment between Venus, Mars, the Moon, **Saturn** and Jupiter, in this order and from west to east. It is certainly an evening to set aside for observing the

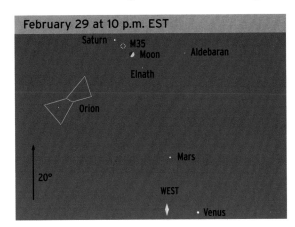

February 29 at 10 p.m. EST

Saturn · ∘ M35
· Moon · Aldebaran
Elnath
Orion

20°

Mars

WEST

Venus

1 Sunday	🌔	
2 Monday	🌔	■ The Moon and Saturn move closer together in the early part of the night (minimum 4.5° at 11 p.m.)
3 Tuesday	🌕	
4 Wednesday	🌕	
5 Thursday	🌕	
6 Friday	🌕	■ 3:50 a.m. Full Moon
7 Saturday	🌕	■ Maximum of Auriga meteor shower (6 meteors an hour)
8 Sunday	🌖	■ The Moon and Jupiter move closer to one another. Visible all night (minimum 3.2° at 9 a.m.)
9 Monday	🌖	
10 Tuesday	🌖	
11 Wednesday	🌗	
12 Thursday	🌗	
13 Friday	🌗	■ 8:38 a.m. Last quarter of the Moon
14 Saturday	🌘	
15 Sunday	🌘	

Red spot rotating
Observed here by the Cassini probe in December 2000, Jupiter's rotation is shown by the movement of its Red Spot.

sky – weather permitting! This is particularly so because the Moon is between the Pleiades and the Hyades, in the constellation of Taurus, one of the regions richest in stars.

There are 29 days in February this year, since 2004 is a **leap year**, and this 29th day offers the possibility of interesting observations with binoculars. At 6 p.m.,

Why are there leap years?

The Earth orbits the Sun in 365.2422 days. It is therefore necessary, every four years, to add a day to the 365 days in the calendar year so that the calendar we use matches the astronomical calendar. This extra day is added in February, which then has 29 days. This is the case in 2004, and again in 2008. The average length of the year is thus extended to 365.25 days. The small remaining discrepancy (about three days in 400 years) is corrected by omitting to add this 29th day to February in some years that would normally be leap years, those that are not divisible by 400, to be precise. Thus 1900, 2100, 2200, and 2300 are not leap years, whereas 1600 and 2000 were. The average length of the year is thus brought back to 365.2425, which represents a discrepancy of only three days in 10,000 years.

the Moon is between the star Elnath, the bull's horn in Taurus, and Saturn. Follow the line that goes from the Moon to Saturn, which are now separated by 9.5°. Halfway between the two stars, you will find a splendid open cluster, Messier 35 (M35). This is a grouping of more than 200 stars situated 2,800 light years away. The stars that make up the grouping are just under 100 million years old. The cluster was discovered by de Chéseaux in 1745, and described by Charles Messier as "a cloud of very small stars near Castor's left foot." Three other open clusters in Messier's catalog can be observed at the same time. On the line joining the Moon and Capella, you will find M36 and M38, M37 being 6° north of the Moon at 6:45 p.m.

Jupiter center stage

Jupiter will be visible as a very bright star that hardly twinkles. If your sight is good, you will not need binoculars to make out the four satellites discovered by Galileo at the beginning of 1610.

These four moons have names that are linked to Jupiter. **Io**, **Europa** and **Callisto** were three of his mistresses, who underwent diverse fortunes and metamorphoses. **Ganymede** was the minion who served nectar to the gods on Olympus. These satellites are very easy to see with binoculars.

Even with a small telescope, it is a very imposing sight: Jupiter seems to be surrounded by his court, which revolves around him in a continuous ballet. Io and Europa will thus be both together on the face of the planet on the 8th at 8:45 p.m., the 15th at 10:40 p.m. and the 22nd at 11:50 p.m. On the evening of the 25th, the four satellites will be on the same side

Jupiter's four large satellites		
	Distance to Jupiter	Length of revolution
Io	422,000 km	1.77 days
Europa	671,000 km	3.55 days
Ganymede	1,070,000 km	7.15 days
Callisto	1,880,000 km	16.69 days

of the planet, and on the morning of the 28th all four will be very near to Jupiter. Jupiter is an enormous planet capable of holding 1,400 Earths. It is the largest of the planets in the solar system, and it is also the one with the largest number of moons: 39 satellites have been counted. Jupiter has been explored by several space probes: Pioneer 10 in 1973, Pioneer 11 in 1974, Voyager 1 and 2 in 1979, Ulysses in 1992, Galileo since 1995 and Cassini in 2000. The surface temperature of this giant of the solar system is of the order of −140°C. The planet's surface is an immense ocean of liquid hydrogen, above which floats a thick atmosphere of hydrogen, helium and methane as well as colloidal dispersions of ammonia compounds and water. The four Galilean satellites are larger than the Moon and, with the exception of Europa, can easily be seen with a small instrument or even with the naked eye by some observers. Ganymede is bigger than Mercury, whereas Callisto is practically the same size as the planet. It was their discovery in

Visibility of Jupiter's Red Spot					
1	12:07 a.m.	12	11:59 p.m.	20	8:35 p.m.
1	7:58 p.m.	13	7:50 p.m.	22	11:13 p.m.
2	9:36 p.m.	15	1:37 a.m.	23	6:04 p.m.
5	11:14 p.m.	15	9:28 p.m.	24	11:51 p.m.
7	7:05 p.m.	17	11:06 p.m.	25	7:42 p.m.
8	9:43 p.m.	18	6:58 p.m.	27	9:20 p.m.
10	10:11 p.m.	20	12:24 a.m.		

16 Monday
- 3 a.m. Moon passes the perigee: 368,320 km

17 Tuesday
- 7:46 p.m. Minimum of Algol

18 Wednesday

19 Thursday

20 Friday
- 4:21 a.m. New Moon
- 4:35 p.m. Minimum of Algol

21 Saturday

22 Sunday

23 Monday
- The Moon and Venus move closer together. Visible in the evening (minimum 3° at 3 p.m.)
- Maximum of the Delta (δ) Leonids meteor shower (five meteors an hour)

24 Tuesday

25 Wednesday
- The Moon and Mars move closer together. Visible in the early part of the night (minimum 0.9° at 9 p.m.)

26 Thursday

27 Friday
- Alignment of Venus, Mars, the Moon, Saturn and Jupiter in the early part of the night
- 10:27 p.m. First quarter of the Moon

28 Saturday
- 6 a.m. Moon passes the apogee: 404,259 km

29 Sunday

1610 that led Galileo to assert that the Earth was not at the center of the Universe, and that it was only one world among others.

Zodiacal light

Of all astronomical phenomena, zodiacal light is the one most easily missed. It was discovered by Gio-Domenico Cassini in 1683 while confirming observations he had conducted in Bologna in 1668. In the evening toward the west, and at the end of the night toward the east, one or two hours after sunset or before sunrise, it is possible to discern a diaphanous glow coming up from the horizon. It is a luminous band about 15° thick, which covers the ecliptic up to about 50°. It is not easy to distinguish from the rays of the setting sun and it is sometimes taken to be the rays of the rising sun just appearing. It has barely any more contrast than the Milky Way, with which it is also sometimes confused. But in fact, zodiacal light is caused by the diffusion of sunlight by particles of dust turning around the Sun in the plane of the ecliptic, where the zodiacal

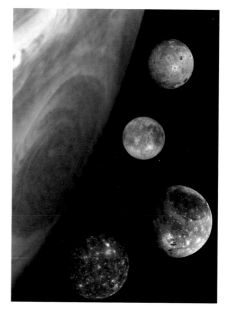

Jupiter and the four Galilean satellites This montage was produced from photos taken by the Galileo probe in 1997.

constellations are found, hence its name. You can appreciate this phenomenon better when the ecliptic is most inclined over the horizon, at intermediate latitudes in February–March and September–October. The inhabitants of tropical regions, for whom the ecliptic is almost perpendicular to the horizon, can see it all year, but only if the atmosphere is very dry.

To study zodiacal light, you need to be under a very dark sky, far away from light pollution, and well accustomed to darkness. You can always see it better by using peripheral vision, that is to say by shifting your gaze from the direction of the zodiac by about 15°. There is another glow, related to zodiacal light, that can be seen more clearly in the middle of the night: this is the Gegenschein, or counterglow. It appears in the form of an elliptical halo of about 10° directly opposite the Sun. Here, again, what is happening is that the Sun is lighting up small dust particles found in the plane of the ecliptic.

Jupiter in figures	
Distance to Sun	5.2 AU, 778,300,000 km
Distance from perihelion	4.951 AU, 740,500,000 km
Distance from aphelion	5.455 AU, 816,600,000 km
Eccentricity of orbit	0.05
Inclination of orbit	1.3°
Equatorial diameter	142,984 km (11.2 times the Earth)
Mass	1.9×10^{27} kg (318 times the Earth)
Revolution period	11.864 years
Orbital velocity	13.06 km/s
Escape velocity	59.6 km/s
Rotation period	9 h 50 min at the equator, 9 h 56 min towards the poles

THE SKY IN FEBRUARY

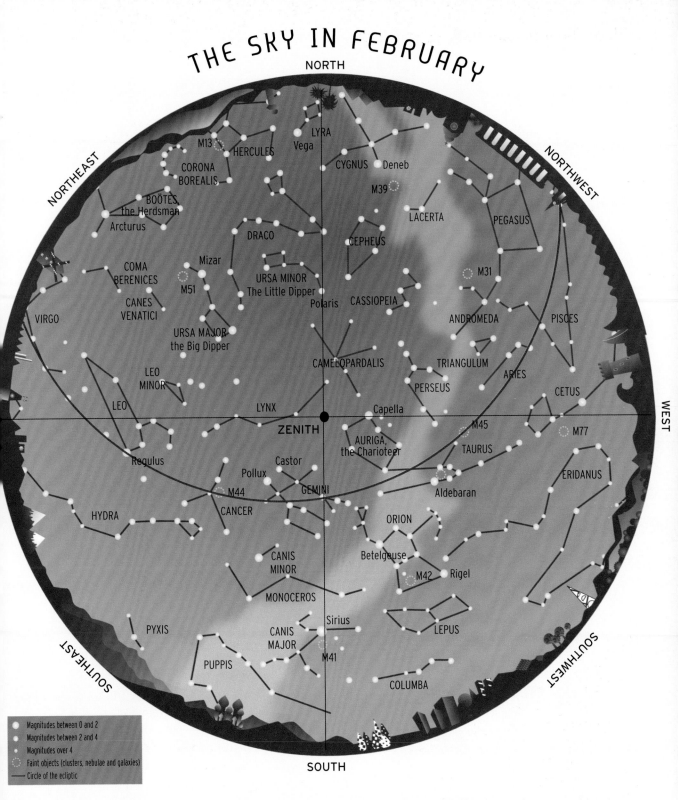

NORTH

NORTHEAST

NORTHWEST

LYRA
Vega

M13 HERCULES

CORONA
BOREALIS

CYGNUS   Deneb

M39

BOÖTES,
the Herdsman

LACERTA

PEGASUS

Arcturus

DRACO

CEPHEUS

COMA
BERENICES

Mizar

M51

URSA MINOR
The Little Dipper

M31

PISCES

CANES
VENATICI

Polaris

CASSIOPEIA

ANDROMEDA

VIRGO

URSA MAJOR
the Big Dipper

CAMELOPARDALIS

TRIANGULUM

ARIES

LEO
MINOR

PERSEUS

CETUS

LEO

LYNX

Capella

M45

WEST

Regulus

ZENITH

AURIGA,
the Charioteer

TAURUS

M77

Castor

Pollux

M44

GEMINI

Aldebaran

ERIDANUS

CANCER

HYDRA

ORION

CANIS
MINOR

Betelgeuse

MONOCEROS

M42   Rigel

Sirius

PYXIS

CANIS
MAJOR

LEPUS

M41

PUPPIS

COLUMBA

SOUTHEAST

SOUTHWEST

SOUTH

- ⚪ Magnitudes between 0 and 2
- ⚫ Magnitudes between 2 and 4
- · Magnitudes over 4
- ⊛ Faint objects (clusters, nebulae and galaxies)
- — Circle of the ecliptic

■ How to use this chart
Hold the chart above your head, matching the word SOUTH that appears at the edge of it with the geographical south of the place you are observing from. Use a compass to help you do this.

■ This chart shows the sky that is visible at a latitude of 45°
If you are further north or further south, Polaris will be higher or lower in the sky

■ Chart of the sky visible at 11 p.m. East
at the beginning of the month;
at 10 p.m. EST in the middle of the month;
at 9 p.m. EST at the end of the month.

A month with your head up in the stars

Shooting stars

The **Auriga** meteor shower visible at the beginning of the month is not very spectacular, with an average of two meteors an hour during the period from the 5th to the 10th, and a maximum of seven. What characterizes the Auriga shower is the slowness of the meteors and the appearance of a few very luminous bolides which have engraved themselves on people's memories, the one on February 27, 1935 (magnitude –5) and especially one on January 31, 1970 (magnitude –9). The **Delta (δ) Leonids**, active from February 5 to March 19, should not be confused with the famous November Leonids. The maximum is only about 20 meteors an hour during the night of the 22nd to the 23rd.

Exploring a region of the sky: Leo, the lion

Among the labors of Hercules, the one best illustrated by the constellations is the episode of the Nemean lion. A lovely Greek legend tells how this lion lived on the Moon and fell down to Earth in the form of a shooting star. The lion ravaged the valley of Nemea until Herakles (the Greek name for Hercules), having tried in vain to kill it with his club and arrows, strangled it with his bare hands. The hero then donned the animal's skin, which no arrow could pierce.

The Sun was in the constellation of Leo 4,000 years ago, near to the star Regulus, at the moment of the summer solstice. It was claimed at the time that the prevailing heat of the season was due to the combined effects of the Sun and the lion. Even today, Corsicans and Italians still use the expression *solleone*, "the Sun in the lion," when talking about very hot weather.

Regulus (or **Alpha (α) Leonis**), a diminutive of the Latin *rex*, "the king," was so named by Nicolaus Copernicus, perhaps in homage to Ptolemy who called it *vasiliskos*, "the ruler of the sky." For the Babylonians, this star was called Lugal, also meaning "the King." Regulus has a magnitude of 1.36, at 30' from the ecliptic, and is 85 light years from the solar system. It is sometimes hidden from view by the Moon and planets. Some 160 times brighter than the Sun, its diameter is five times bigger than that of our planet and its surface temperature reaches 13,000 K.

Two fairly bright stars appear through binoculars 5° to the west of Regulus: they are **Leonis 18** and **19**, whose respective magnitudes are 5.6 and 6.4, the first being red and the second yellow.

R Leonis is a long-period variable (310 days) whose magnitude goes from 5 to 10. It is a red giant, situated about 600 light years away, and its brightness is 200 times greater than that of the Sun. The next maxima of R Leonis will take place in September 2003 and July 2004.

Denebola (or **Beta (β) Leonis**), an abbreviation of the Arabic *al dhanab al asad*, "the lion's tail," is a star quite similar to Sirius but situated 43 light years away, hence its lesser magnitude of 2.14.

Algieba (or **Gamma (γ) Leonis**), probably a corrupt form of *al jabbah*, "the

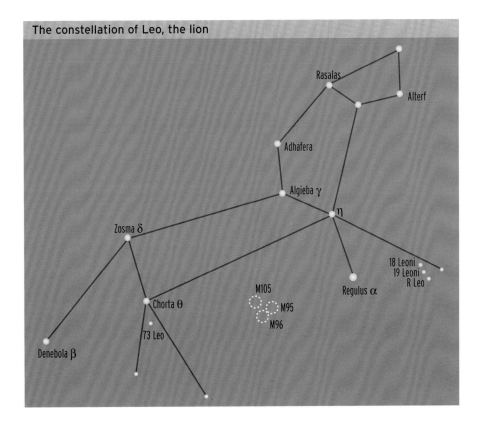

The constellation of Leo, the lion

mane," is a very beautiful double star discovered by William Herschel in 1782. Its period is very long, being six or seven centuries. The two components have magnitudes of 2.14 and 3.39, Algieba attaining 1.98 overall. Close to this star can be seen the radiant of the Leonids meteor shower, the Leonids being at their maximum around November 17. **Zosma** (or **Delta** (δ) **Leonis**), probably derives from the Greek *zosma*, meaning "belt." Zosma is about 80 light years away, and its apparent magnitude is 2.55. **M65** and **M66** are two beautiful spiral galaxies (NGC 3623 and NGC 3627) discovered by Pierre Méchain in March 1780. They are 38 million light years away and probably belong to the Virgo Cluster. On a line joining Beta

Leonis and Alpha Leonis is a small star, **73 Leo**, visible to the naked eye and practically in a direct line from Theta. M65 and M66 are located to the east of this star.

M95 and **M96** are two other galaxies discovered by Pierre Méchain in March 1781. Less than 1° to the north of M96 is NGC 3379, sometimes called **M105**, together with its two companions NGC 3384 and NGC 3389. The galaxies in this group, about 29 million light years distant from Earth, interact with one another. To find M95 and M96, you need to look again at the line joining Beta and Alpha Leonis. M105 is one-third of the way along this line from Alpha. M95 and M96 are 1° south of M105.

Practical Astronomy

How to use the Internet

The Internet offers extraordinary possibilities for the amateur astronomer, containing both information about current experiments in space and the results of observations carried out by the most powerful telescopes. Numerous amateur astronomers maintain their own sites and publish their work. Observatories throughout the world present their research and keep you informed of opportunities for visits by the public. Last but not least, the Net is a wonderful tool for guiding and facilitating any visits to the sky you may want to undertake.

Observation advice

Here are a few important sites that are mostly free. Occasionally, however, there is a charge for access to more professional information, like receiving announcements of discoveries from the International Astronomical Union or using certain databases. This usually takes the form of an annual subscription.

www.imcce.fr

The site of the French Institut de Mécanique Céleste et de Calcul des Éphémérides provides numerous ephemerides for objects in the solar system, and the rising and setting of stars, whatever the observation point. It is possible to calculate observation conditions for eclipses of the Sun and Moon from any position.

www.heavens-above.com

This site provides possibilities for a huge variety of calculations, in particular for ephemerides of artificial satellites. It is kept well up to date and enables you to calculate, from the place where you are, visibility from the International Space Station. In addition, it tells you when the most spectacular satellites are going to be passing across the sky, and gives information about iridium flashes (from orbiting satellites), some of which are visible in broad daylight.

www.stargazing.net/astrotips

In English or Portuguese, this site provides different programs of astronomical

Image of galaxy
You can explore the galaxies on the ESO website: this is Centaurus A

Triton
This satellite of Neptune, observed by the Voyager probe, is an example of the pictures available on the NASA website.

calculations of all kinds and for different purposes: ephemerides, cartography, for planetariums, adjusting telescopes, making photographic or electronic settings. The programs offered are downloadable or can be used on line.

Specialized journals also have sites where information required for observation is regularly updated. A few of these include:

www.cieletespace.fr

skyandtelescope.com

www.astronomy.com

Information which is virtually live

During the observation of comets, their orbital characteristics change and become clearer. The Internet offers a means of using this information with an almost live update, enabling you to calculate their positions.

cfa-www.harvard.edu/icq/icq.html

for example, gives you everything on observable comets, both past and future.

One network also gives information on probable polar lights, and forecasts maxima for meteor showers:

www.spaceweather.com

Another site offering to alert Web-surfers by e-mail of the probability of polar lights is:

www.dcs.lancs.ac.uk/iono/aurorawatch

Certain sites even offer to help with the working out of personalized ephemerides using powerful means of calculation. The International Astronomical Union site

cfa-www.harvard.edu/cfa/ps/mpc.html

gives the latest developments regarding small objects in the solar system and the requirements and priorities for observation.

Space live on the Net

Thanks to certain sites, like those of the ESA and NASA, it is possible to follow the progress of space missions live. In addition to the pictures obtained by SOHO, information on the progress of the Cassini-Huygens probe, and the positions of all the probes at present operating in interplanetary space, is kept permanently up to date.

www.eso.org

This remarkable site of the European Southern Observatory presents images taken from ESO telescopes, among which is the VLT.

www.esa.int

The site of the European Space Agency enables you to follow missions in progress. In particular, you will find on it live pictures of the Sun taken by the SOHO probe.

www.nasa.gov

This site gives the latest information about missions in progress and offers numerous links enabling you to access the results of past missions. You will also find on it details of space projects and the calendar for launches. An extraordinary collection of pictures of the planets can be viewed on photojournal.jpl.nasa.gov and on www.stsci.edu - the site of the Hubble space telescope - which publishes remarkable pictures gathered by the telescope.

Volcanic eruption on Io, to be seen on the NASA website.

Introduction to astronomy

The four gaseous planets
These giants are, in order of their distance from the Sun: Jupiter, Saturn, Uranus and Neptune.

Discovering gaseous planets

More than 750 million kilometers from the Sun, and taking 12 years to revolve around it, is the impressive planet Jupiter, the first of the four gaseous planets in the solar system. At such a distance away, the Sun appears no bigger than a very large, bright star. Jupiter, Saturn, Uranus and Neptune were formed in a cold environment, which was rich in ice crystals and gas molecules. This abundance of volatile elements was the reason for their reaching such colossal sizes, far larger than the telluric planets near to the Sun.

Jupiter

Jupiter measures 142,884 kilometers at the equator, almost 12 times the Earth's diameter. Its rapid rotation, in less than 10 hours, means that it is substantially flattened out at the poles. Physical models suggest that an inorganic core the size of the Earth lies at the center of the planet, where there is a pressure of 100 million bars and a temperature of 30,000 K. This core is surrounded by a thick layer of liquid helium, then a coat of liquid hydrogen that stretches out to 40,000 kilometers from the center.

This hydrogen is in metallic form: the electrons are pulled out of it and their circulation generates a powerful magnetic field. Above this extends a layer of neutral hydrogen (molecular), which becomes gaseous on the surface. In this atmosphere, clouds of ice crystals and methane circulate, gathering together in belts parallel to the equator. The atmosphere is agitated by violent winds (at more than 300 km/h), containing huge cyclones and anticyclones, like the famous Red Spot, which is more than 20,000 kilometers across.

Jupiter was flown over by the Pioneer 10 and 11 probes (1973–74), Voyager 1 and 2 (1979), Galileo, which was put into orbit in December 1995, and Cassini in December 2000. In addition to the discovery of a thin ring of dust, situated between 58,000 and 64,000 kilometers from Jupiter, these probes photographed its multiple moons in detail. Jupiter possesses no fewer than 28 natural satellites. Most of them are small asteroids which have been captured, several tens of kilometers in diameter, but four are of a size comparable with our Moon: Io, Europa, Ganymede and Callisto. Pulled at by Jupiter's powerful gravitational field, Io experiences phenomenal internal heating, causing permanent volcanic eruption. It would appear that Europa, on the other hand, conceals a subterranean ocean under its icy crust.

Saturn

Twice as far from the Sun as Jupiter (1.5 billion kilometers), Saturn takes 29 years to complete its orbit, surrounded by a retinue of rings and satellites. As with Jupiter, its rapid rotation causes its flattening out at the poles. It is also made up of hydrogen and helium but, being less solid and therefore less tightly packed than Jupiter, its overall density does not exceed 0.7 – in an ocean which corresponded to its size, Saturn would float. The few spots visible on its surface show that its atmosphere is also stirred up by violent winds, reaching 1,500 km/h at its equator.

Saturn possesses 30 moons, but only Titan is of any size: 5,150 kilometers in diameter, it is surrounded by a thick atmosphere of nitrogen. Tethys, Dione, Rhea, and Iapetus measure between 1,000 and 1,500 kilometers and all appear to consist mainly of ice.

But Saturn is above all known for its magnificent ring system. Made up of innumerable particles in orbit, these rings are divided into about half a dozen "tracks" which begin 7,000 kilometers above the atmosphere and extend up to 480,000 kilometers from the planet. The innermost ring, known as Crêpe, seems to

Saturn
Saturn and three of its satellites are observed here through the Hubble Space Telescope.

Uranus
Uranus' ring system and 9 of its 17 satellites are observed here through the Hubble Space Telescope.

Uranus

The third giant planet, Uranus, orbiting almost three billion kilometers from the Sun and hardly visible to the naked eye, was discovered by the astronomer William Herschel in 1781. It is peculiar in having a rotational axis tilted at 98° in relation to the plane of its orbit: instead of being "upright," Uranus lies on its "side." This surprising situation remains unexplained – perhaps a collision with a protoplanet, early in its history, was responsible for it. The internal composition of Uranus is differentiated from that of Jupiter and Saturn. Its rocky core is thought to be wrapped in a covering of water ice, ammonia and methane. On the surface, the atmosphere of hydrogen and helium is colored blue by traces of methane.

Uranus possesses a system of a dozen rings of barely a few tens of kilometers across. Five moons were known prior to the space age: Oberon and Titania (1,500 to 1,600 kilometers in diameter), Ariel and Umbriel (1,100 to 1,200 kilometers), and Miranda (480 kilometers). The Hubble telescope and the Voyager 2 probe counted 16 additional moons, measuring between 25 and 150 kilometers in diameter.

be made of 2-meter blocks, whereas the main ring, or B ring, is made up of smaller particles (from 10 centimeters to 1 meter). Then follows a dark gap, called the Cassini Division, beyond which the A ring includes particles of varying size. Other thin rings (F, G and E) complete the system. The rings, made up of ice and inorganic dust, are not very thick, a few tens of meters on average. It is therefore not surprising that they disappear completely when viewed edge on.

The gaseous planets in figures					
	Jupiter	Saturn	Uranus	Neptune	Pluto
Average distance from sun (millions km)	778	1,429	2,875	4,504	5,900
Revolution period (years)	11.9	29.5	84	164.8	247.7
Equatorial diameter (km)	142,984	120,536	51,108	50,538	2,350
Rotation period	9h 50m	10h 39m	17h 14m	16h 3m	6d 9h 17m
Density	1.31	0.69	1.30	1.64	1.99
Gravity (Earth = 1)	2.54	1.07	0.8	1.2	0.01

Neptune

Neptune is the fourth and the most distant of the gaseous planets: its existence was deduced mathematically from the disturbances it was causing in the orbit of Uranus. Thanks to the calculations of Urbain Le Verrier in 1845, it was located by the astronomer Johann Galle on September 23, 1846.

Neptune makes an almost circular orbit in 165 years, 4.5 billion kilometers from the Sun. With a diameter of 50,538 kilometers at the equator, it is of comparable size to Uranus and its internal composition is probably identical. Its atmosphere of hydrogen is also tinted blue by methane. When the Voyager 2 probe flew over it in August 1989, it detected layers of cloud and giant anticyclones, recalling Jupiter's great Red Spot.

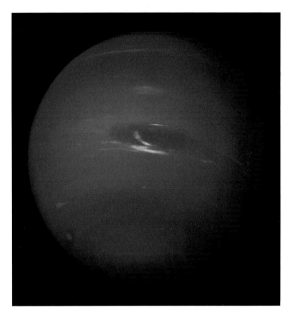

Neptune
Neptune's great equatorial cyclone is clearly seen on this photo taken by the Voyager 2 probe in August 1989.

Pluto, the exception

Discovered by the astronomer Clyde Tombaugh in 1930, Pluto is the last planet in the solar system. It is not a giant ball of hydrogen and helium, but a small rocky body 2,324 kilometers in diameter, smaller than our Moon. In 1978, the astronomer James Christy, taking advantage of the fact that Pluto had moved closer in relative terms, discovered a moon 1,270 kilometers in diameter, which he named Charon. Pluto is probably made up of 70% rock and 30% ice. Its surface appears to be covered with methane ice and nitrogen. The Hubble Telescope was able to make out just a few spots on the surface, dark at the equator and bright at the poles.

Neptune also possesses a system of fine rings, discovered by the astronomer André Brahic when occultation by a star occurred in May 1981. Their existence was confirmed by Voyager 2. The outer ring is divided into three individual bulges or "arcs," which André Brahic named "Liberty, Equality, Fraternity" as a tribute to the bicentenary of the French Revolution. Neptune possesses eight moons, of which only one, Triton (2,705 kilometers in diameter) is of a respectable size. When Voyager 2 flew over it, it was able to observe calderas, mounds and other manifestations of "volcanic" eruptions of soft ice on the surface. It also photographed two geysers, rising to a height of 8,000 meters. These phenomena are, in all probability, due to the sublimation ▶ of frozen nitrogen under Triton's surface. Our exploration of the solar system will always be full of surprises!

DEFINITION ▶
Sublimation: the direct change from a solid to a gaseous state (without passing through a liquid stage).

Cassini-Huygens en route for Saturn

An extraordinary space experiment is under way. The Cassini craft left Earth on October 15, 1997, and is moving toward Saturn, which it will reach in July 2004. Cassini is carrying with it a European probe called Huygens, which will explore the atmosphere of Titan, the largest of Saturn's satellites, on January 14, 2005.

Cassini-Huygens
These probes, seen here as models, are en route for Saturn, which they will reach in July 2004.

Saturn is the last of the planets visible to the naked eye. It is so far away that light takes more than an hour to travel to Earth, but this journey has already been made by three space probes: Pioneer 11 flew over Saturn on September 1, 1979, Voyager 1 on November 12, 1980, and Voyager 2 on August 26, 1981. Cassini will thus reach Saturn more than 23 years after the last flight.

Space billiards

Cassini is named in honor of Gio-Domenico Cassini who, between 1672 and 1686, discovered four satellites of Saturn and the division between the rings that now bears his name. The mass of the vehicle is almost two metric tons; when Huygens' 350 kilograms is added, together with the fuel, there is a total launch mass of 5.6 metric tons – more than is possible, with current technology, to propel directly from the Earth as far as Saturn. Consequently, it was planned that the probe would fly by several planets in succession to "steal" a little of their orbital speed.

The difference in mass between the planet and the probe is such that the exchange of energy entails a slowing down of the former by a few centimeters per billion years, and an acceleration of the latter by a few kilometers per second.

The Cassini-Huygens space probe thus began its journey by heading toward the Sun. It was twice accelerated by Venus before flying around the Earth in August 1999 to pick up speed. It could be said that it took almost two years to gather momentum, and that it really left Earth in that month of August 1999. Its speed was then 19 km/s, whereas the speed provided by the launch rocket Titan, the most powerful available at the time, was only 11 km/s.

Cassini passed close to Jupiter on December 30, 2000, to gain another 16 km/s cruising speed and fly off toward Saturn.

Huygens
This is how Huyghens' landing on Titan soil (forecast for January 14, 2005) should look.

Other aspects of the journey

The results of the investigation of Jupiter using the Cassini probe have been used in a joint operation with the Galileo probe, which has been exploring the giant planet since 1995. The Cassini probe was tested in the process, and all the experiments with imagery, radar and surveillance of plasma and cosmic dust showed that the craft's instruments were functioning normally. It was thus possible to obtain new knowledge and a new collection of pictures of Jupiter's atmosphere.

After the mission to Titan, scheduled to be conducted on January 14, 2005, Cassini will continue to explore the system of Saturn, traveling until June 30, 2008, after which date it is expected that its mission will comto an end.

The exploration of Titan

The Cassini probe is American, but Huygens is European. It was named in homage to Christiaan Huygens who, in 1655, discovered that Saturn's ring did not touch the planet and that it possessed a satellite, Titan.

After separating from Cassini, Huygens will enter Titan's atmosphere in January 2005, and will be slowed down by its heat shield before a variety of parachutes are deployed. These will control its two-hour descent, after which a soft landing will allow ground measurements to be carried out over several minutes. These measurements will be transmitted to Earth via Cassini.

Titan is of considerable interest, being one of the largest satellites in the solar system and possessing a complex atmosphere, rich in organic compounds, which must resemble the Earth's atmosphere at an early stage in its history. But the temperature is so low, just 90 K, that any development of life has been rendered impossible. Being able to enter this atmosphere and study will be the same as having access to the conditions that preceded the appearance of life on Earth. It will seem like defrosting the matter that existed on our planet just before the first forms of life sprang up.

The atmosphere of Titan, one of Saturn's satellites

March

Beauty of the night
The comet Hyakutake, as seen in March 1996, from near Nice in southern France. The straight tail of gas is quite clearly defined, whereas the tail of dust, which is more diffuse, is difficult to make out.

RENDEZVOUS IN THE SKY

1 ■ The Moon and Saturn move closer together in the early part of the night
6 ■ The Moon and Jupiter move closer together throughout the night
24 ■ The Moon and Venus move closer together in the early part of the night
25 ■ The Moon and Mars move closer together in the early part of the night

A month with your head up in the stars

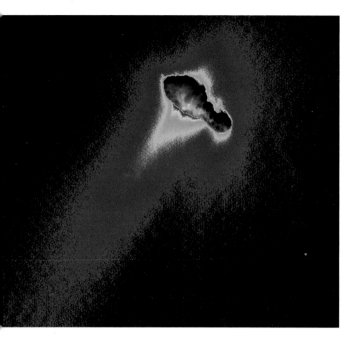

March is the month of the spring equinox, which this year takes place on the 20th. This month of transition between winter and summer is characterized by the opposition of Jupiter, which rises at the moment when the Sun is setting and remains visible all night.

Borrelly's comet seen from Deep Space 1
The nucleus, which is only a few kilometers, is surrounded by an immense halo of several hundred thousand kilometers.

Observations of the month

Taurus, Orion and the Winter Triangle are beginning their descent toward the west. Capella, Castor, Pollux, Procyon and Sirius dominate at the beginning of the night, then Leo, together with Regulus, take over the sky. In the northeast, Ursa Major is moving upwards, vertically, followed by the Corona Borealis. Spring is announced by the arrival of Spica, in the constellation of Virgo, and Arcturus, in Boötes.

The planets

The month begins with **Saturn** and the Moon moving toward one another (4.5°). The Moon then passes near to Pollux, in Gemini (2° on the evening of the 2nd). On the 5th, the moon

All times are given in Eastern Standard Time (EST). Calculations have been made for a latitude of 44° north and a longitude of 80° west.

TIMES OF RISING AND SETTING FOR THE SUN, MOON AND FIVE PLANETS VISIBLE WITH THE NAKED EYE														
Date	Sun		Moon		Mercury		Venus		Mars		Jupiter		Saturn	
	Rises	Sets	Rises	Sets	Rises	Sets	Rises	Sets	Rises	Sets	Rises	Sets	Rises	Sets
3/1	6:57 a.m.	6:08 p.m.	12:24 p.m.	4:05 a.m.	7:02 a.m.	5:54 p.m.	8:28 a.m.	10:00 p.m.	9:23 a.m.	11:56 p.m.	6:16 p.m.	7:18 a.m.	12:32 p.m.	3:49 a.m.
3/5	6:51 a.m.	6:13 p.m.	4:49 p.m.	6:36 a.m.	7:02 a.m.	6:19 p.m.	8:21 a.m.	10:09 p.m.	9:14 a.m.	11:54 p.m.	5:57 p.m.	7:02 a.m.	12:16 p.m.	3:33 a.m.
3/10	6:42 a.m.	6:19 p.m.	9:46 p.m.	8:28 a.m.	7:00 a.m.	6:51 p.m.	8:12 a.m.	10:20 p.m.	9:03 a.m.	11:51 p.m.	5:34 p.m.	6:41 a.m.	11:57 a.m.	3:14 a.m.
3/15	6:33 a.m.	6:25 p.m.	3:54 a.m.	12:16 p.m.	6:57 a.m.	7:24 p.m.	8:04 a.m.	10:30 p.m.	8:53 a.m.	11:48 p.m.	5:11 p.m.	6:20 a.m.	11:37 a.m.	2:54 a.m.
3/20	6:24 a.m.	6:32 p.m.	6:36 a.m.	6:22 p.m.	6:52 a.m.	7:54 p.m.	7:56 a.m.	10:40 p.m.	8:43 a.m.	11:46 p.m.	4:48 p.m.	5:59 a.m.	11:14 a.m.	2:31 a.m.
3/25	6:15 a.m.	6:38 p.m.	8:22 a.m.	10:50 p.m.	6:44 a.m.	8:17 p.m.	7:48 a.m.	10:49 p.m.	8:34 a.m.	11:43 p.m.	4:26 p.m.	5:38 a.m.	10:55 a.m.	2:12 a.m.
3/30	6:06 a.m.	6:44 p.m.	12:09 p.m.	3:32 a.m.	6:33 a.m.	8:28 p.m.	7:41 a.m.	10:58 p.m.	8:24 a.m.	11:40 p.m.	4:03 p.m.	5:17 a.m.	10:36 a.m.	1:53 a.m.

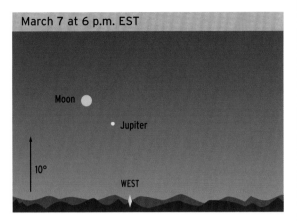

March 7 at 6 p.m. EST

Moon

· Jupiter

10°

WEST

passes above Pollux to approach **Jupiter**. It is interesting to observe the Moon's movement in the vicinity of a bright star like Pollux: watch the Pollux–Jupiter–Moon triangle as it becomes deformed between the 5th and the 6th, and try to work out the Moon's angular velocity. On the 9th, the Moon rises near Spica, one of the important stars of the zodiac, in the constellation of Virgo. Nearly two weeks later, when it is a very fine crescent, the Moon will pass near to **Mercury** on the evening of the 22nd (5°). The show will continue as it moves nearer to **Venus**, which is easier to observe, on the 24th (2°), then toward Mars (0.25°) during the night of the 25th to 26th, and finally, as it is in its first quarter, toward Saturn (4°) on the 28th. Jupiter, although clearly present

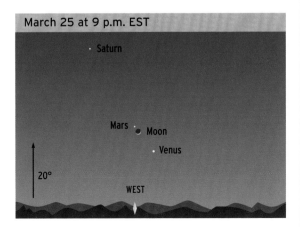

March 25 at 9 p.m. EST

· Saturn

Mars · Moon
· Venus

20°

WEST

1 Monday ■ The Moon and Saturn move closer together. Can be seen during the first part of the night (minimum 4.5° at 5 a.m.)

2 Tuesday

3 Wednesday

4 Thursday ■ 12 midnight Jupiter in opposition

5 Friday

6 Saturday ■ The Moon and Jupiter move closer together. Visible all night (minimum 3.1° at 11 a.m.)
■ 6:17 p.m. Full moon

7 Sunday

8 Monday

9 Tuesday

10 Wednesday

11 Thursday ■ 6:20 p.m. Minimum of Algol
■ 11 p.m. Moon passes the perigee: 369,511 km

12 Friday

13 Saturday ■ 4 p.m. Last quarter of the Moon

14 Sunday ■ 3:09 p.m. Minimum of Algol

15 Monday

Equinox and refraction

The word "equinox" comes from the Latin aequinoctium and means that day and night are of equal duration. However, we can see that this is not true. The spring equinox takes place on the 20th at 1:50 a.m. On that day, the Sun rises at 6:50 a.m. and sets at 6:58 p.m. The length of time when the Sun is up is therefore 12 hours 8 minutes, and the night lasts for 11 hours 52 minutes. Why is there this discrepancy of 16 minutes when day and night should be perfectly equal? The cause of the mystery comes from the Earth's atmosphere: this spherical layer of air deforms images, and this is known as "refraction." The bigger the layer of air, the bigger the refraction is. When the Sun is on the horizon, either at sunrise or at sunset, its low light cuts through the thickest blanket of air and its image is pulled up by 0.5°, which corresponds to the maximum refraction. Each day, it moves back 1°, to go round the sky in a year. Between rising and setting, it recedes by 0.5° angular distance, which it covers in two minutes. The length of time it stays light is therefore increased by two minutes. In total, refraction is responsible for increasing the presence of the Sun above the horizon by four minutes and its daily movement by two minutes, making six extra minutes. Another two minutes are added because, this year, the equinox takes place at 1 a.m. and not at 12 noon. This is why night and day are not perfectly equal, even at the moment of the equinox.

in the evening sky at the end of this month, will not be approached by the Moon until April 2. During these evenings, it is very easy to read the line of the ecliptic in the sky and to notice that the planets move among the stars only in a very clearly defined band around this line.

Shooting stars

Two meteor showers appear in March. The Camelopardalids are slow, penetrating the atmosphere at barely 10 km/s and they seem to move much less rapidly than other meteors. The maximum, which is difficult to specify because the number of meteors is small, takes place on the 22nd. The March Geminids are more numerous: about 40 meteors an hour, on the same day as the Camelopardalids. They are fairly slow and appear to be moving southwards.

An area of the sky to expore: Boötes, the Herdsman

This star formation (or asterism), representing the figure of a herdsman, has been influenced by its neighbor, Ursa Major. Initially, it seems that the Greeks saw it as a herdsman guiding his cart. With the Romans, the designation became more explicable on account of the seven threshing oxen (*septem triones*, septentrion) that they regarded as making up Ursa Major. In reality, these interpretations apply to the main star, Arcturus, probably one of the very first stars to have been given a name. The name Arcturus derives from the Greek *arctouros*, the "leader" or "keeper" of the bear. Arcturus does indeed follow Ursa Major constellation as if he were guiding or watching over it. Sometimes the constellation is also imagined to be a plowman pushing his plow (in Britain, the constellation is often called the Plow). Since the herdsman invented the plow to make men's work simpler, he was put into the sky by Jupiter at the insistence of Ceres, the goddess of agriculture.

This constellation has also carried the name of Icarius, whom Dionysus taught to make wine and was murdered by his friends who thought he had poisoned them, whereas they had simply become drunk. Icarius's dogs, Asterion and Chara, found their master's body and lay down to die beside him. This explains the presence of the Hunting Dogs constellation, Canes Venatici, near to the Herdsman.

Arcturus (or **Alpha** (α) **Boötis**), the Bear Keeper or Herdsman, is the fourth star of the sky in terms of brightness (magnitude –0.06), and is situated 37 light years away from us. Its diameter is about 25 times that of the Sun but it is 115 times brighter and has a mass equivalent to four times that of the Sun. It is therefore a giant of very low density. Its surface temperature is of the order of 4,200 K, giving it a golden yellow color. This star is moving and is at present passing close to the solar system. It was much further away 500,000 years ago and invisible to the naked eye. It will continue to move nearer for another few thousand years, and then move away to become invisible once more, 500,000 years later. Arcturus is a Population II star – that is to say it belongs to the elliptical halo surrounding the galaxy, a relic of the latter's primitive form, which contains old stars, low in heavy elements.

Mirac (or **Izar**, or **Epsilon** (ε) **Boötis**), is one of the most beautiful double stars that can be observed with a medium-sized telescope. It is a good test for a 2-inch (50 mm) diameter lens. The two components, one yellow and the other

16 Tuesday

17 Wednesday

18 Thursday

19 Friday

20 Saturday
■ 1:50 a.m. Spring equinox
■ 5:45 p.m. New Moon

21 Sunday

22 Monday
■ Maximum of Camelopardalids (2 meteors an hour) and March Geminids (30 meteors an hour) meteor showers

23 Tuesday

24 Wednesday
■ The Moon and Venus move closer together. Visible in the early part of the night (minimum 2.2° at 4 p.m.)

25 Thursday
■ The Moon and Mars move closer together. Visible in the early part of the night (minimum 0.25° at 8:10 p.m.)

26 Friday

27 Saturday
■ 2 a.m. The Moon passes the apogee: 404,520 km

28 Sunday
■ 6:53 p.m. First quarter of the Moon

29 Monday
■ 7 a.m. Greatest easterly elongation of Mercury (18.9°)
■ 10 a.m. Greatest easterly elongation of Venus (46°)

30 Tuesday

31 Wednesday

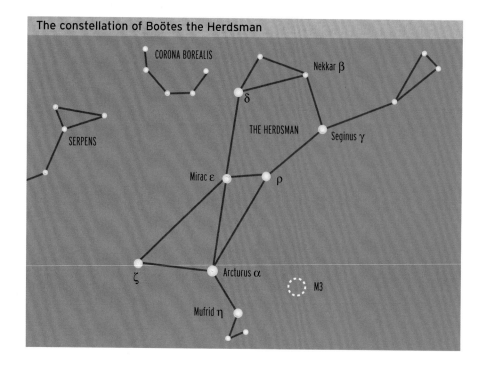

The constellation of Boötes the Herdsman

CORONA BOREALIS

Nekkar β

δ

THE HERDSMAN

Seginus γ

SERPENS

Mirac ε

ρ

ζ

Arcturus α

M3

Mufrid η

blue, with magnitudes of 2.5 and 4.9, are separated by 2.8".

Zeta (ζ) Boötis is an interesting double star discovered by William Herschel in 1780, and one of the nearest doubles to the solar system, at a distance of barely 22 light years. The two components, which are yellow and reddish purple, have magnitudes of 4.7 and 6.9 respectively, and are at present separated by 6.6". The period of the system is 150 years.

The Messier marathon

It is in this month of March, toward the equinox, when theoretically it is possible to observe all the objects in Charles Messier's catalogue in a single night. Messier was the astronomer at the Paris Observatory from 1758 to 1784. Put in charge of finding the comet whose return had been forecast by Edmund Halley about 50 years earlier, he made a systematic exploration of the sky and

encountered diffuse objects liable to be confused with comets because of their appearance, but which turned out to be fixed and permanent in the cosmos. He began to draw up a list of them. The first one he entered in his catalogue, Messier 1 or M1, is the Crab Nebula, in the constellation of Taurus. With the help of Pierre Méchain, he counted 104 objects; a few more were added later. Today, Messier's catalogue contains 110 numbers, but there are a few errors, like M91 and M102, which do not correspond to any diffuse object, or M40, which is in fact a double star in Ursa Major.

A game played by some experienced amateur astronomers consists of training their telescopes on the maximum number of objects in Messier's catalogue in a single night, from the moment the Sun sets in the west, over a 12-hour period. The marathon starts by searching

THE SKY IN MARCH

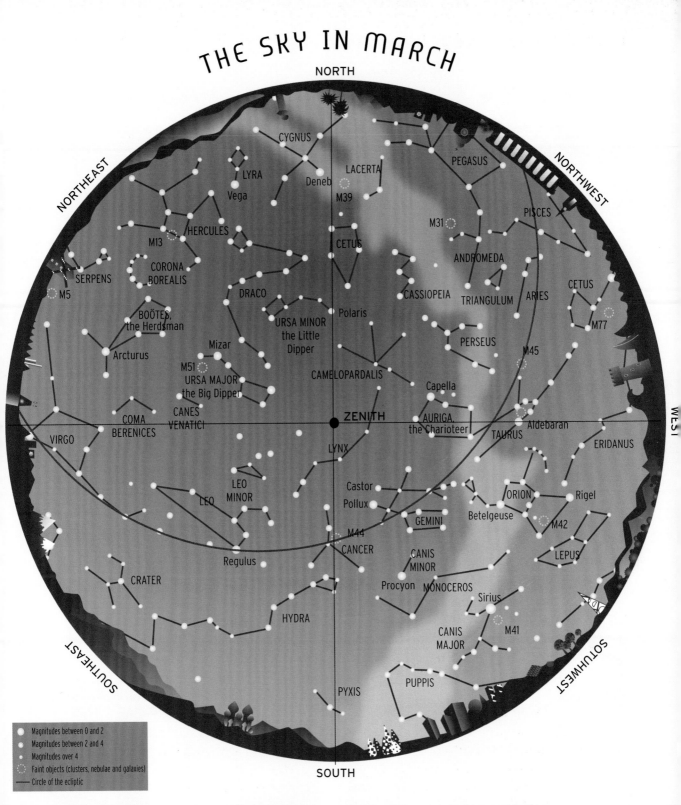

NORTH

CYGNUS
LACERTA
PEGASUS
NORTHWEST
LYRA
Deneb
M39
Vega
M31
PISCES
NORTHEAST
HERCULES
M13
CETUS
ANDROMEDA
CETUS
CORONA
BOREALIS
SERPENS
CASSIOPEIA
TRIANGULUM
ARIES
M77
M5
DRACO
Polaris
BOÖTES,
the Herdsman
URSA MINOR
the Little
Dipper
PERSEUS
M45
Mizar
Arcturus
CAMELOPARDALIS
M51
URSA MAJOR
the Big Dipper
Capella
WEST
CANES
VENATICI
AURIGA,
the Charioteer
ZENITH
COMA
BERENICES
Aldebaran
VIRGO
LYNX
TAURUS
ERIDANUS
LEO
MINOR
Castor
LEO
Pollux
ORION
Rigel
GEMINI
Betelgeuse
M42
M44
CANCER
LEPUS
Regulus
CANIS
MINOR
CRATER
Procyon
MONOCEROS
Sirius
HYDRA
CANIS
MAJOR
M41
SOUTHEAST
PYXIS
PUPPIS
SOUTHWEST

SOUTH

- ○ Magnitudes between 0 and 2
- ○ Magnitudes between 2 and 4
- · Magnitudes over 4
- ○ Faint objects (clusters, nebulae and galaxies)
- — Circle of the ecliptic

■ **How to use this chart**
Hold the chart above your head, matching the word SOUTH that appears at the edge of it with the geographical south of the place you are observing from. Use a compass to help you do this.

■ **This chart shows the sky that is visible at a latitude of 45°**
If you are further north or further south, Polaris will be higher or lower in the sky.

■ **Chart of the sky visible at 11 p.m. EST** at the beginning of the month; at 10 p.m. EST in the middle of the month; at 9 p.m. EST at the end of the month.

M1
The Crab Nebula is the remains of a supernova in the constellation of Taurus

out **M77** in Cetus, the Whale, and finishes with the observation of **M30**, in the constellation of Capricornus.

Clearly it is not possible to carry out real observations at such a pace. It is more to provide a sense of achievement. However, in the knowledge that all the objects in the catalogue are accessible during the nights of March, it is fascinating to select a few and indulge in a sort of astronomical tourism.

All the objects mentioned here are observable with 2.5-inch (60 mm) telescopes, but their structure is only revealed if this diameter is increased. This is why Messier, who used small optical equipment, could not distinguish

between globular clusters, planetary nebulae and galaxies. Messier's principal objects can be located with the help of the monthly charts set out in this book.

Globular clusters are spectacular objects: the most amazing ones are **M3** in Canes Venatici, **M13** and **M92** in Hercules, **M4** in Scorpius and **M5** in Serpens. With slightly more powerful instruments than Messier's, about 8 inches (200 mm) in diameter for example, you can begin to "resolve" some of these clusters – in other words, you can see clearly that you are looking at an enormous concentration of stars and not a gas cloud.

It is also interesting to look for planetary nebulae in Messier's catalogue.

M51 and M4
The Whirlpool Nebula, in the center, is a spiral galaxy in Canes Venatici. M4, on the right, is a globular cluster in Scorpius.

M31
The Andromeda Nebula is a spiral galaxy.

M42
The Orion Nebula is a region where stars are formed.

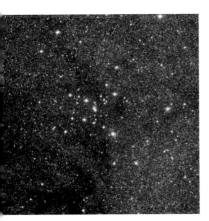

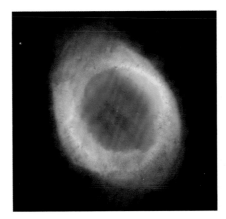

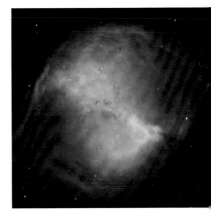

These are attractive to look at, relatively circular and symmetrical, sometimes widely spread out, sometimes not so much. A telescope 8 inches (200 mm) in diameter allows you clearly to make out the structure of most of these objects, the show-stealers being **M57, Lyra's ring nebula, M27, the Dumbbell**, in Vulpecula (the Fox), and **M97**, in Ursa Major.

The locations of new star formation are also present in Messier's catalogue, and some of them look impressive through binoculars: for example, **M8** and **M20**, called **the Lagoon** and **the Trifid** respectively, which are in Sagittarius, as well as **M42, the Orion nebula**.

The open clusters, which are spectacular when viewed through simple binoculars, are mainly **M44, the Beehive cluster**, in Cancer, and **M45, the Pleiades cluster**, in Taurus.

Finally, there are numerous galaxies in Messier's catalogue. Sometimes visible with the naked eye, like M31, most only through binoculars – but it requires an instrument of more than 6 inches (150 mm) in diameter to be able really to appreciate their character and beauty. The "easiest" are **M31, the great Andromeda galaxy, M33** in the Triangulum, **M51**, in Canes Venatici, and **M81** and **M82**, in Ursa Major. **M95** and **M96**, in Leo, are also accessible.

Practical astronomy

How to position an observation instrument

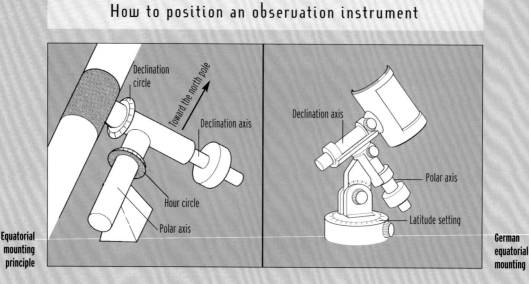

Declination circle

Toward the north pole

Declination axis

Hour circle

Polar axis

Declination axis

Polar axis

Latitude setting

Equatorial mounting principle

German equatorial mounting

To observe the stars or take photographs of them, you need to set up your equipment, that is to say position it correctly and adjust it so as to train it on and follow the objects whose apparent movement is caused by the Earth's rotation.

With an altazimuth mounting

For a simple instrument like a beginner's telescope, positioning is easy. Since one of the axes it moves around is vertical and the other is horizontal, you just need to position the mounting as vertically as possible with the help of a plumb line, and this will enable you to observe without difficulty and even carry out measurements.

For instruments which have computer-aided guidance or which use the "Go-To" system of automatic training, the main axis must be vertical. Then you need to direct the instrument toward the north, horizontally, to give the mounting its reference position. A compass is usually sufficient, although magnetic north is separated by several degrees from geographical north. These systems require the introduction of position coordinates, i.e., latitude and longitude, which you can calculate with the help of a detailed map. The machine then trains itself automatically on a bright star. You re-center this star with the help of the instrument's memory functions. After validation, you aim at a second bright star, wait for the centering, and then validate. The instrument is now operational, calculating and pointing out for you the position of the several thousand objects it has in its memory. It can assess the positions of the planets, the Sun, the Moon and certain asteroids.

Top-of-the-range instruments, fitted with **GPS** (global positioning system) and automatic compass, set themselves up without the observer having to intervene, on condition that the main axis has been vertically positioned beforehand. They locate the position coordinates thanks to the network satellites, set themselves up entirely automatically and point directly at the objects of the observer's choice.

With an equatorial mounting

An equatorial mounting moves around two axes: the polar axis, which is parallel to the Earth's rotational axis, and the declination axis, which is at right angles to the polar axis.

On most equatorial mountings, the polar axis is adjustable by latitude, which is usually graduated. You therefore need to begin by making the angle of the axis the same latitude as the observation point, for example 37.5° in San Francisco. Next, you point the axis more or less toward Polaris. By locking the telescope at the 90° declination, you move the foot so as to bring Polaris into the center of the field. When this operation is complete, the basic position is set up. You can improve the position by pointing at the celestial north pole, separated by 0.6° from Polaris in the direction of Beta (β) Ursae Minoris.

The quick positioning of equipment

One method has proved very useful for observers who have to travel to a site where they want to work and would like to avoid wasting too much time at night making adjustments to their telescope. It is possible make such adjustments while still daylight, if you know or can calculate the declination of the Sun on the observation date. Set your telescope firmly on this declination on the day when you want to observe, and also fix your polar axis at the latitude of your location. Once you have made these two settings, leave them alone and aim at the Sun (using the instrument's shadow to avoid any risks associated with viewing the Sun through a lens), just altering the polar and vertical axes of your mounting. When the Sun is coming through the instrument, your positioning is complete. You may possibly need to make finer adjustments later, once night falls, but you will already have solved the greater part of the problem.

For photographs or closer investigation of objects, there are motorized equatorial mountings driven by a clockwork movement.

Prices of automatic mountings

Automated altazimuth mountings fit telescopes of all sizes. Prices vary between about $700 for a small telescope 2.5 inches (60 mm) in diameter and $4,000 for an 8-inch (200 mm) telescope, the price of the instrument together with its mounting. An 11-inch (280 mm) telescope fitted with a "Go-To" system and automatic positioning by GPS costs around $6,500 or roughly the price of the same equipment installed on a good quality equatorial mounting.

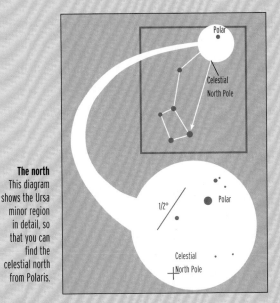

The north
This diagram shows the Ursa minor region in detail, so that you can find the celestial north from Polaris.

Introduction to astronomy

The Hale-Bopp comet
Hale-Bopp was one of the most spectacular visitors from space in the last few decades.

Discovering comets

From time immemorial, comets have worried people. Appearing suddenly, often looking strange, and then disappearing without warning, they could not be classified with the other objects in the night sky. It was not until the 17th century that these wondrous objects were finally put into the category of celestial mechanics and recognized as heavenly bodies that turned around the Sun. Nowadays, it is thought that comets are frozen balls of dust, several tens of kilometers in diameter, revolving around the Sun in very large numbers and at an enormous distance. Gravitational disturbances sometimes cause these far-off objects to become displaced and hurled toward the Sun, which has the effect of warming them and making them partially melt. The tails of comets are made up of materials, both solid and

gaseous, which were released when fusion occurred and are now slowly dispersing in interplanetary space.

The origin of comets

The solar system was formed in an immense cloud of hydrogen and helium accompanied by a few traces of heavier elements. This cloud contracted under the effect of gravitation and a multitude of secondary concentrations of matter came into being. These agglomerations gave birth to the planets, which were relatively small lumps of matter, rich in hydrogen, helium, oxygen, carbonates and silicates, and which continued to move within this developing planetary system. All these objects, subject to universal gravitation, were more or less distributed in the equatorial plane of the emerging Sun, just as small satellites are

The Hale-Bopp comet
These two photos were taken in March 1997 in the south of France.

distributed in ring structures around the large planets. When thermonuclear reactions were set off at the center of the Sun, the force of the blast propelled most of the small bodies in the Sun's vicinity toward the far reaches of the solar system. It was these objects that formed the Oort Cloud, containing several hundred million comet nuclei and spreading out to between 1,000 and 100,000 AU. The total mass of comet matter in the cloud is estimated to be about 1,000 times that of Earth. The extremities of the Oort Cloud are found on the edges of the solar system, almost half the distance separating us from the nearest stars. The comets found in this coment take several million years to turn around the Sun, and they are therefore defined as long-period comets ▶.

The concentrations of matter that were further away from the Sun were not affected by the solar reaction blast and remained in the place where they had taken form, in the plane of the ecliptic. These are the comets of the Kuiper Belt, which contains several billion comets and spreads out to between 30 and 1,000 AU. From time to time a very long-period comet appears to burst out of this cloud and enter the central regions of the system, where the planets are found. Comets of this type have shorter periods than those emanating from the Oort Cloud.

Morphology of a comet

From the observational point of view, a comet is composed of a nucleus, which is always invisible, surrounded by a fairly extensive head or tail. If the nucleus measures several kilometers, the tail can easily spread out over thousands of kilometers. The tail's appearance

▶ **DEFINITION**
The period of a comet is the time between two of its perihelions (the point in its orbit when the comet comes nearest to the Sun) - that is to say the time taken for it to orbit the Sun completely.

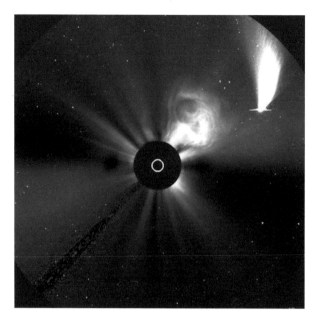

Dive towards the Sun
The NEAT comet is seen here to the right of the disk masking the Sun, being disturbed by a large solar eruption.

changes over time, which means that regular observation needs to be carried out. In fact, the comet has two tails, each very different from the other. The gas tail is composed of very light atoms and molecules from the nucleus, driven back by the light and particles from the Sun. This is why it is straight and always in the opposite direction from the Sun. It is often blue, the color resulting from the atoms whipped up by the solar wind; it may measure several hundred million kilometers and sweep across several planets. The second tail is shorter and more diffuse, curved toward the Sun and composed essentially of dust. It is this dust, emanating from the nucleus, which, when it meets the Earth, causes the phenomenon of meteors or shooting stars.

Voyages of the comets

It sometimes happens that comets crash into planets. In the case of Jupiter, for example, this must occur about once a century. An impact on Jupiter, very similar to that of Comet Shoemaker-Levy 9 in 1994, was observed by the astronomer Cassini in 1690. As for the Sun, it is hit by several tens of comets per year; these are mostly detected thanks to pictures taken by the SOHO probe, which has been observing the Sun uninterruptedly since early 1996. The Earth can also be hit by comets. The last spectacular incident occurred on June 30, 1908, in Siberia, near the river Tunguska, causing the destruction of hundreds of square kilometers of forest, which were fortunately uninhabited.

The periods of comets are very variable. The shortest is that of Encke, which is 3.3 years. The longest are measured in millions of years, a length of time which is not, of course, significant. Spectacular periodic comets are quite rare, the most famous being Halley's Comet, which returns about every 76 years and will again be visible from Earth in 2061. The comet Hale-Bopp

Comet Hale-Bopp
In this photo taken in 1997, the comet's two tails, one of gas and the other of dust, are clearly visible.

illuminated the sky in the spring of 1997. This is a comet whose long period changed during its last journey through the central zone of the solar system, from 4,206 to 2,380 years, mainly on account of the gravitational disturbances caused by Jupiter.

Comets bear the name of the person who discovers them, with the notable exception of a few famous comets, such as Encke or Halley, which bear the name of the scientist who calculated their orbit.

Evening companions

A few comets pass close to the Earth. The one that has come nearest to us is Lexell, which rubbed shoulders with our planet at about 3.5 million kilometers on July 1, 1770. Halley's Comet has passed three times within close proximity: 13 million kilometers on April 1,374, 14 million on April 19, 607, and 5 million on April 10, 837. The next time it will pass close to us will be May 7, 2134: its distance will then be only 14 million kilometers, and it will be visible in full daylight. Spectacular comets that are visible to the naked eye are quite rare, about a dozen per century. But more and more comets are being discovered by automatic detectors and spacecraft, particularly by the SOHO probe, in the immediate vicinity of the Sun; the number of SOHO's new finds reached about a hundred in 2001, while in the same year, the LINEAR (Lincoln Near Earth Asteroid Research) program, which automatically detects small bodies in the solar system, revealed about 20. None of these is visible to the naked eye. However, most of the population of the solar system is made up of comets; billions of them are revolving around the Sun, and a few hundred move near to the central region of the system each year. Each century, about half a dozen pass less than 15 million kilometers away from the Earth. When will the next one be?

Halley's Comet
In 1986, the Giotto probe managed to get close to the comet and photograph its nucleus.

Operation Comet!

Research into comets has progressed considerably thanks to space missions, which have brought back decisive information about their origin and composition. The ambition now is to approach and even to reach these comets, and gather more precise data concerning their matter, which originated with the first moments of the solar system.

The first mission to a comet was conducted by the ICE (International Cometary Explorer) probe, launched in August 1978. Two comets were observed: Giacobini-Zinner on September 11, 1985, and Halley, from a distance of 28 million kilometers, in March 1986.

Two other missions followed, destined for Halley. These were Vega 1 and 2,

launched on December 15 and 20, 1984, which passed 39,000 kilometers away on March 6, 1986 (Vega 1), and 8,000 kilometers away on March 9 (Vega 2), producing the first picture of a comet nucleus. Two Japanese probes, Sakigake and Suisei, were also sent toward Halley in 1985.

The European probe, Giotto, which was launched on July 2, 1985, passed 600 kilometers away from Halley's nucleus on March 13, 1986. Despite the enormous risks of the probe being destroyed by the cometary dust encountered at 70 km/s, the experiment was a complete success. The only problem encountered was when a particle hit a deflector, depriving the astronomers of pictures during the stage when the probe was moving away from

Halley's Comet
Photographed here in 1985, when it last passed across our sky. The next time will be 2061.

The Deep Space Impact probe

Stardust, which was launched on February 7, 1999, will pass 150 kilometers away from the comet Wild 2 on January 1, 2004, and in January 2006 will bring back a capsule of specimens of cometary matter gathered from around the comet. Comet Nucleus Tour (CONTOUR), launched on August 13, 2001, will reach the comets Encke on November 12, 2003, Schwassmann-Wachmann 3 on June 18, 2006, and d'Arrest on August 16, 2008.

Deep Space Impact will be launched on January 6, 2004. This is made up of two spacecraft, which will meet up with Comet Temple 1 on July 4, 2005. The smaller of the two probes will crash into the bright side of the comet's nucleus in order to make a crater 120 meters in diameter and 25 meters deep. Rosetta is a European mission which aims to send a probe into the vicinity of a comet and deposit a landing craft, called Roland, to perform a remarkable feat for the very first time: it will land on a comet and proceed to actually analyze the surface matter of this distant object. Rosetta was to have been launched in January 2003 toward Comet Virtanen, but the setbacks with the Ariane 5 rocket have led to the mission being postponed until 2004 or 2005. This will be to a new target, which remains to be chosen.

the comet. The probe was then put into terrestrial orbit and sent back toward the comet Grigg-Skjellerup, which it reached on July 10, 1992.

Deep Space 1, launched on October 24, 1998, passed 26 kilometers away from the asteroid Braille (9989) on July 28, 1999, before moving on to Comet Borelly on September 22, 2001. The moment of the rendezvous with Borelly was observed through the Hubble Space Telescope.

Progress in the means of observation

Remarkable observations can be made with space telescopes, which detect not only what is visible, but also wavelengths that are invisible on Earth. It is thanks to their use that our knowledge of these stars has progressed. Indeed, some comets are discovered from space, either accidentally by telescopes exploring the infrared part of the spectrum, or during observations of the environment immediately surrounding the Sun, into which a good number of comets are precipitated and detected at the moment of their disappearance. The SOHO solar observatory satellite enabled the discovery of several hundred comets just at the moment when they were falling into the Sun.

The Rosetta probe with Roland, its landing craft.

April

Distant cluster
The cluster's gravitational field alters the path of light coming from more distant galaxies and intensifies it, deforming and multiplying their images. The Hubble Telescope reveals objects situated 2.2 billion light years away.

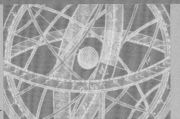

A month with your head up in the stars

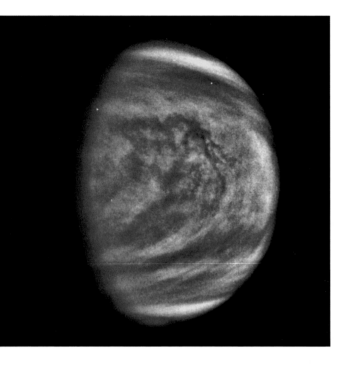

At the beginning of the month, Jupiter, Regulus and the Moon rise early in the evening. The five planets visible to the naked eye together with the Moon – six out of the seven "moving bodies" – are then above the horizon. But the star of the month is Venus, resplendent on the western horizon, performing a spectacular ballet with Mars and the Moon at the end of the month to light up our spring evenings.

Observations of the month

Venus
Beautiful Venus hides under a thick layer of cloud.

Take advantage of these lovely spring evenings to observe Algol's final minima before this star, representing the eye of the Medusa, disappears in the Sun's light. The evening of the 5th will be marked by the Moon passing very near (0.75°) to the star Gamma (γ) Virginis, also known as

Porrima. It is one of the most beautiful double stars you can see, the two components being practically identical, with a magnitude of 3.65. The two components of the system are at present moving toward one another. Minimal separation (0.4") will be reached in 2007. The period of this pair, situated

All times are given in Eastern Standard Time (EST) until April 4, when the change to Daylight Saving Time takes place. After that they are given in Eastern Daylight Time (EDT). Calculations have been made for a northern latitude of 44° and a western longitude of 80°.

Date	Sun Rises	Sun Sets	Moon Rises	Moon Sets	Mercury Rises	Mercury Sets	Venus Rises	Venus Sets	Mars Rises	Mars Sets	Jupiter Rises	Jupiter Sets	Saturn Rises	Saturn Sets
TIMES OF RISING AND SETTING FOR THE SUN, MOON AND FIVE PLANETS VISIBLE WITH THE NAKED EYE
4/1	6:02 a.m.	6:46 p.m.	1:27 p.m.	3:36 a.m.	6:27 a.m.	8:27 p.m.	7:38 a.m.	10:58 p.m.	8:21 a.m.	11:38 p.m.	3:50 p.m.	5:07 a.m.	10:29 a.m.	1:49 a.m.
4/5	6:55 a.m.	7:51 p.m.	7:25 p.m.	6:07 a.m.	7:14 a.m.	9:22 p.m.	8:32 a.m.	12:04 a.m.	9:14 a.m.	12:35 a.m.	4:32 p.m.	5:51 a.m.	11:14 a.m.	2:34 a.m.
4/10	6:46 a.m.	7:55 p.m.	12:45 a.m.	9:03 a.m.	6:53 a.m.	9:00 p.m.	8:26 a.m.	12:09 a.m.	9:06 a.m.	12:32 a.m.	4:10 p.m.	5:30 a.m.	10:56 a.m.	2:15 a.m.
4/15	6:38 a.m.	8:01 p.m.	4:21 a.m.	3:01 p.m.	6:32 a.m.	8:25 p.m.	8:19 a.m.	12:13 a.m.	8:58 a.m.	12:28 a.m.	3:49 p.m.	5:10 a.m.	10:37 a.m.	1:57 a.m.
4/20	6:29 a.m.	8:07 p.m.	6:00 a.m.	7:28 p.m.	6:11 a.m.	8:33 p.m.	8:13 a.m.	12:15 a.m.	8:50 a.m.	12:24 a.m.	3:28 p.m.	4:49 a.m.	10:19 a.m.	1:39 a.m.
4/25	6:21 a.m.	8:13 p.m.	8:59 a.m.	12:41 a.m.	5:52 a.m.	7:56 p.m.	8:25 a.m.	12:14 a.m.	8:43 a.m.	12:19 a.m.	3:07 p.m.	4:29 a.m.	10:01 a.m.	1:21 a.m.
4/30	6:14 a.m.	8:19 p.m.	2:28 p.m.	3:26 a.m.	5:36 a.m.	6:32 p.m.	7:57 a.m.	12:10 a.m.	8:36 a.m.	12:14 a.m.	2:47 p.m.	4:09 a.m.	9:44 a.m.	1:03 a.m.

April 3 at 4 a.m. EST

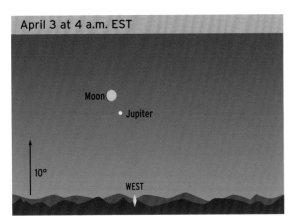

Moon • Jupiter

10°

WEST

33 light years away, is 171 years and the maximum distance between the stars can reach 6". On the 6th, the Moon will pass near to Spica (3°) (the name refers to the ear of corn held by the Virgin). Spica is the fourteenth star in order of apparent brightness. It could be considered as the prototype of the star with a magnitude of 1. It is 2,300 times more luminous than the Sun but is 262 light years away.

The planets

During the night of the 2nd to the 3rd, the meeting between **Jupiter** and the Moon (3°) underlines the presence of six moving stars above the horizon in the early evening, with **Mercury** due east, then **Venus**, **Mars** and **Saturn** to the south.

April 23 at 10 p.m. EDT

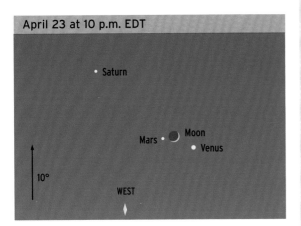

• Saturn

Mars • • Moon
• Venus

10°

WEST

1 Thursday	■ Maximum of the meteor shower Tau (τ) Draconides (3 meteors an hour)
2 Friday	■ The Moon and Jupiter move closer together. Visible in the early part of the night (minimum 2.8° at 2 p.m. EST)
3 Saturday	■ 4:54 p.m. Minimum of Algol
4 Sunday	■ Daylight Saving Time begins, Eastern Daylight Time (EDT = UT - 4)
5 Monday	■ 7:05 a.m. Full Moon
6 Tuesday	
7 Wednesday	■ 10 p.m. Moon passes the perigee: 364,548 km
8 Thursday	
9 Friday	
10 Saturday	
11 Sunday	■ 11:47 p.m. Last quarter of the Moon
12 Monday	
13 Tuesday	
14 Wednesday	
15 Thursday	

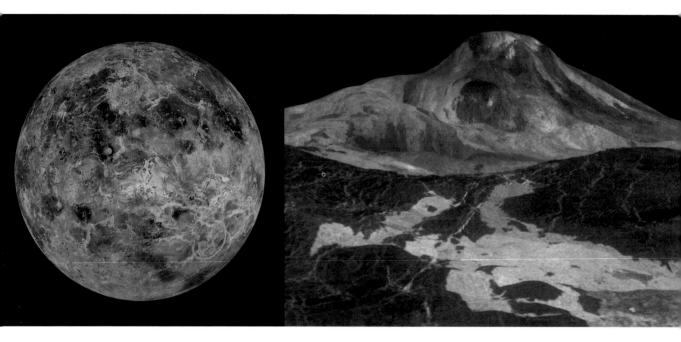

Venus revealed
It has been possible to reconstruct the surface of Venus by radar, with the help of the Arecibo radio telescope and the Magellan, Pioneer nd Venera probes.

If the eastern and western horizons are nice and clear, you can continue to see the six moving bodies, in the early evenings, until the 4th. The Moon is crossing the constellation of Virgo between the 4th and the 7th. And, from the 23rd onward, there is a superb show in the evenings, with Venus, Mars and the Moon center stage, watched over by Saturn on the western horizon. A lovely right-angled triangle is formed by the Moon, Mars and Venus on the 23rd at 6 p.m., then, at 6 p.m. on the 24th, a near-perfect parallelogram can be seen in outline, formed by Saturn, the Moon, Venus and Mars. This will be followed by the Moon and Pollux (2°) moving closer together during the evening of the 26th.

Venus center stage

An inferior planet, that is to say orbiting between the Sun and the Earth, Venus is never very far from the Sun. The maximum angular separation that it can reach with the Sun is 47°. It is therefore not possible to see it in the middle of the night, only in the evening, toward the west, when it follows the setting Sun, or in the morning, toward the east, where it rises before the Sun. Its visibility before and during sunset has earned it the name of "evening star." Because of an ancient legend in the French region of Provence involving Maguelone, daughter of the king of Naples, and Pierre of Provence, Venus is also known there as "Maguelone," "the beautiful Maguelone who runs after Pierre of Provence (Saturn) and marries him every seven years" (Alphonse Daudet). In very good weather conditions, you can see the planet in broad daylight with the aid of a small optical instrument. It is even possible to see it with the naked eye in the middle of the day, but only at high altitude, where the atmosphere is less dense and the sky is darker.

Again with the aid of small instruments, it is possible to study the phases of Venus. It was, in fact, during just such an observation that Galileo, between February and

Giant volcano
The surface of the planet Venus is revealed by radar observations conducted from space. This volcano is one of the most impressive in the solar system.

Venus
The clouds of Venus are observed here in ultraviolet by Mariner 10. On the right: Like our Moon, Venus undergoes phases.

October 1610, discovered that Venus had very similar characteristics to the Moon – that it, together with the other planets, was naturally dark and that we only see the parts illuminated by the Sun.

You can also quite easily observe the apparent diameter of Venus, which is evolving considerably. It varies from 9.6" at the superior conjunction (maximal distance), to 65.4" at the inferior conjunction (minimal distance). With an average magnitude of –4.4 at maximal elongation, Venus is the most brilliant "star" in the sky after the Sun and the Moon. In this month of April, its magnitude is –3.8, the apparent diameter going from 13.6" to 12". Venus is able to pass in front of the solar disk, but these transits are rare. They occur four times, alternately in June and December, according to a complex 243-year cycle: 122

Venus in figures	
Distance to Sun	108,600,000 km
Distance from perihelion	107,480,000 km
Distance from aphelion	108,940,000 km
Eccentricity of orbit	0.007
Inclination of orbit	3.39°
Equatorial diameter	12,103.6 km
Mass	4.8685×10^{24} kg
Revolution period	224.695 days
Orbital velocity	35.02 km/s
Escape velocity	10.36 km/s
Rotation period	243 days (retrograde)

16 Friday

17 Saturday

18 Sunday

19 Monday
- 9:24 a.m. New Moon
- 9:39 a.m. Partial eclipse of the Sun visible in southern Africa

20 Tuesday

21 Wednesday
- Maximum of the Lyrids meteor shower (10 meteors an hour)

22 Thursday

23 Friday
- The Moon moves closer to Venus (minimum 1.5° at 6 a.m.) and closer to Mars (minimum 2.2° at 5 p.m.). Visible early in the night
- 8 p.m. Moon passes the apogee: 405,403 km

24 Saturday

25 Sunday

26 Monday

27 Tuesday
- 1:39 p.m. First quarter of the Moon

28 Wednesday

29 Thursday
- The Moon moves closer to Jupiter. Visible all night (minimum 3.7° at 10 p.m.)

30 Friday

years, 8 years, 105 years, 8 years. This exceptional phenomenon will occur in the month of June next, on the 8th.

Shooting stars

There are numerous shooting stars in April, especially the **Lyrids**, with a maximum on the 22nd. The Lyrids were observed by the Chinese from the 7th century BC onwards. It is thought they are the remnants of Comet Thatcher, whose period is 450 years and whose last passage was in 1861. You will also be able to observe the maximum of **Tau** (τ) **Draconides** on the 1st, as well as several showers which seem to come from the constellation of Virgo and whose maxima are between the 7th and the 20th. But April is characterized by bolides. The "April bolides," which do not have a defined radiant, illuminate the sky with their spectacular trails and frequently cause falls of meteorites.

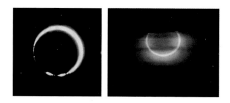

Annular eclipses of the Sun
The lunar disk does not mask the whole of the solar disk.

Eclipse of the Sun in southern Africa

A partial eclipse of the Sun, on the 19th, heralds the total eclipse of the Moon on May 4. The annular eclipse of the Sun will be visible in southern Africa, from the Cape from 12:52 p.m. to 3:22 p.m., in Johannesburg from 1:25 p.m. to 3:35 p.m., in Pretoria from 1:27 p.m. to 3:36 p.m., and at Windhoek, in Namibia, from 1:23 p.m. to 2:31 p.m. (all times UTC).

To see the next total eclipse of the Sun on the North American continent, we will have to wait until August 21, 2017.

A region of the sky to explore: Corona and Serpens

In Western culture, the beautiful constellation of Corona Borealis is linked to Ariadne, who symbolizes the three stages of love in a woman's life: first she is initiator, then abandoned lover, and finally divine spouse. In love with Theseus, Ariadne enabled him to escape from the labyrinth thanks to the ball of thread given to her by Daedalus. She was soon abandoned by Theseus, and taken to Olympus by Dionysus, who gave her a crown made by Hephaestus. It was this crown that was placed in the sky to perpetuate the memory of Ariadne. In Classical times, it was simply called the Crown or Corona: the epithet "borealis" was added when another constellation of the Corona was created in the southern hemisphere. In our own day,

Ariadne (in its French version, Ariane) is the name of the European space launcher. The program, begun in 1972, enabled Europe to escape from a veritable labyrinth of negotiations and troubles of all kinds in which it had got bogged down, and to acquire independence in the space arena. The name brought good luck since the program has now been "crowned" with success!

Gemma (or **Alpha** (α) **Coronae Borealis**), "stone" or "pearl" in Latin, is sometimes called Alphecca, from the Arabic *al naïr al fakkah*, "the diamond of the dish." It is also known as Margarita Coronae, the "pearl in the crown," which, in some cultures, including the French region of Provence, has turned into Sainte

THE SKY IN APRIL

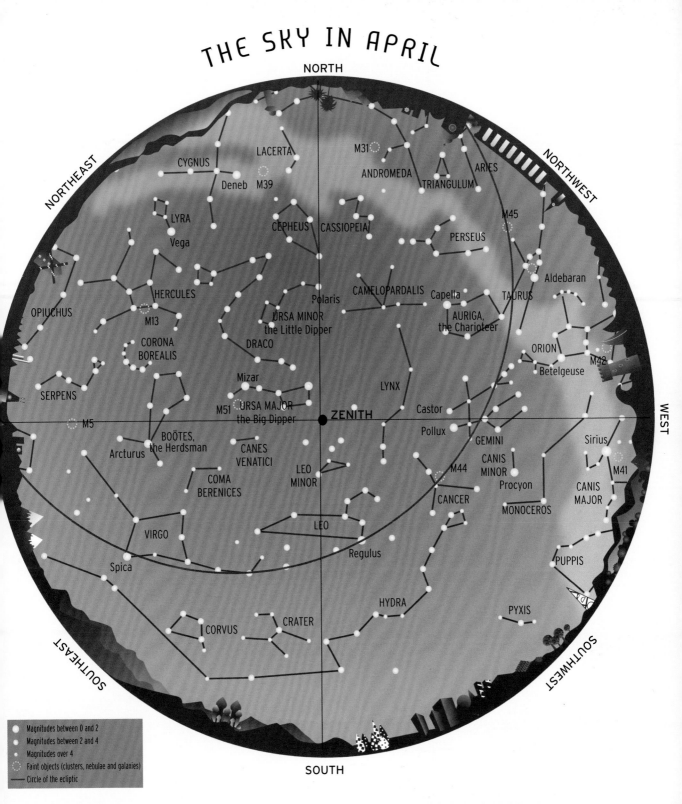

NORTH

NORTHEAST

NORTHWEST

CYGNUS
LACERTA
M31
ANDROMEDA
ARIES
TRIANGULUM
Deneb M39
LYRA
Vega
CEPHEUS
CASSIOPEIA
PERSEUS
M45
HERCULES
CAMELOPARDALIS
Capella
TAURUS
Aldebaran
OPIUCHUS
M13
Polaris
AURIGA,
the Charioteer
CORONA
BOREALIS
URSA MINOR
the Little Dipper
DRACO
ORION
M42
Betelgeuse
SERPENS
Mizar
LYNX
M5
M51
URSA MAJOR
the Big Dipper
Castor
WEST
ZENITH
Pollux
Sirius
BOÖTES,
the Herdsman
Arcturus
CANES
VENATICI
LEO
MINOR
GEMINI
CANIS
MINOR
M41
COMA
BERENICES
M44
Procyon
CANIS
MAJOR
VIRGO
LEO
CANCER
MONOCEROS
Spica
Regulus
PUPPIS
HYDRA
PYXIS
CORVUS
CRATER

SOUTH

SOUTHEAST

SOUTHWEST

- Magnitudes between 0 and 2
- Magnitudes between 2 and 4
- Magnitudes over 4
- Faint objects (clusters, nebulae and galaxies)
— Circle of the ecliptic

■ How to use this chart
Hold the chart above your head, matching the word SOUTH that appears at the edge of it with the geographical south of the place you are observing from. Use a compass to help you do this.

■ This chart shows the sky that is visible at a latitude of 45°
If you are further north or further south, Polaris will be higher or lower in the sky.

■ Chart of the sky visible at 11 p.m. EDT
at the beginning of the month;
at 10 p.m. EDT in the middle of the month;
at 9 p.m. EDT at the end of the month.

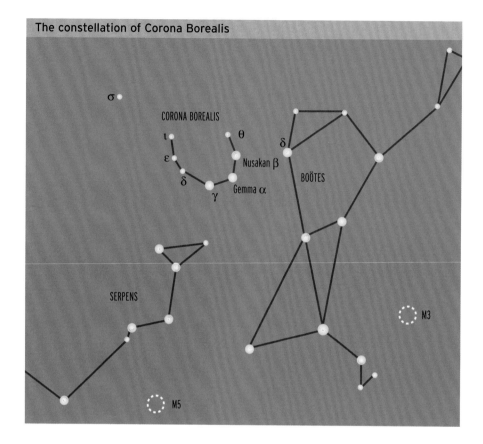

The constellation of Corona Borealis

Marguerite! Gemma is 71 light years away from Earth and is a hot star with a magnitude of 2.22. It is also a double spectroscopic star with a period of 17.35 days.

Nusakan (or **Beta** (β) **Coronae Borealis**), is situated about 100 light years away. It is another double spectroscopic star (with a period of 10.49 days), probably a member of the Hyades cluster.

Sigma (σ) **Coronae Borealis** is a lovely double star discovered by Struve in 1827. Its period, which is still imprecise, is somewhere between 340 and 1,600 years. The distance between the two stars has been increasing since the time of their discovery, and they are now separated by 7.1". The main component, of a magnitude of 5.6, is itself a double spectroscopic

with a period approaching eight days. The distance of this system from the Earth is thought to be 70 light years.

Tau (τ) **Coronae Borealis** is a recurrent nova with a magnitude varying between 10 and 2. Its last maximum took place in 1946, and the one before that in 1866. Its distance from the Earth is estimated at 2,600 light years. Apart from RS Ophiuchi, it is the only nova of this type that is periodically visible to the naked eye.

The Serpens constellation

Just under the Corona, we find the Serpent's Head. The serpent in question is Asclepius, the Roman god of medicine, represented by the large neighboring constellation of Ophiuchus, also known as the Serpent-bearer. Serpens is the only

The constellation of Serpens

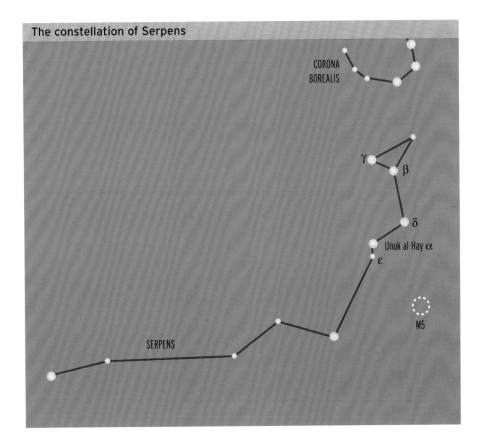

CORONA
BOREALIS

γ

β

δ

Unuk al Hay α

ε

M5

SERPENS

constellation divided into two distinct parts: the Serpent's Head, to the west of Ophiuchus, and the Serpent's Tail, to the east of Ophiuchus.

Unuk al Hay or **Unukalhai** (otherwise known as **Alpha** (α) **Serpentis**), "the serpent's neck," has a magnitude of 2.65. This star has a companion with a magnitude of 12 at 58". The brighter one is pale yellow in color, while its companion is blue.

M5 is one of three beautiful globular clusters in the northern sky, together with M13 in Hercules and M3 in Canes Venatici. M5 was discovered in 1702 by Gottfried Kirch and cataloged by Messier in 1764. It was described as a "cosmic snowball," because it stands out so much against the background of the sky. Through binoculars, a hazy star with a

magnitude of 7 can be seen. Through a small telescope, however, you can clearly see a bright nebula, and with an aperture of more than 4 inches (100 mm), it is possible to resolve the cluster into stars. You can then make out its slightly elliptical shape. M5 is 24,500 light years away and is moving away from the solar system at 50 km/s. Its diameter is 140 light years.

M5
This is one of the most beautiful globular clusters in the sky.

Instrument of the month: the Dobson's telescope

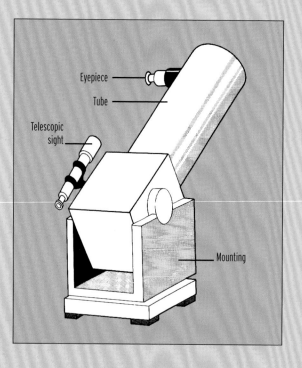

Eyepiece

Tube

Telescopic sight

Mounting

Dobsonian telescope
The simplicity of this telescope puts large diameters within reach of amateurs.

A few decades ago, in California, amateur astronomers were attempting to revolutionize the idea of telescopes by finding very cheap optical and mechanical solutions. John Dobson suggested creating mirrors which were slimmer than the standard Pyrex ones; in the opinion of experts at the time, it was essential for the thickness of the mirror to be at least equal to ⅙ of its diameter. Dobson's mirrors were made from pieces of glass, which were often round, porthole-type windows, propped up by a blanket folded to form a cushion. He discovered tricks for making telescopes from planks of wood and invented a way of making the mechanism both precise and flexible. In the 1970s, he started to demonstrate his enormous telescopes at meetings of amateur astronomers,

proving that it was possible for them to use equipment which up until then had been inaccessible, and, what was more, for the price of instruments with a diameter two or three times smaller.

Large diameters accessible to all
Numerous amateurs took up Dobson's ideas, and quickly created their own large-diameter telescopes measuring between 8 and 32 inches (200 and 800 mm). As with all Newtonian telescopes, since the eyepiece is at the head of the tube, the instrument must not be too long; this will ensure that you can continue to focus comfortably. A classic opening ratio for this type of telescope is on the order of 6 or 7 – that is to say, that the tube is 6 or 7 times longer than the diameter of the mirror.

The thickness of Dobson's mirrors is on the order of ⅟₁₆ of the diameter. The mounting is a simple cube, open at one side and at the top. This cube pivots around its vertical axis, ensuring that the instrument moves in all directions. The tube rests, via two couplings 6 inches (15 cm) in diameter, on two half-bearings hollowed out of the box's sides. These bearings define the horizontal axis around which the telescope can be directed. Flexibility of movement is ensured thanks to Teflon sill plates. This basic type of mounting enables you to have easy access to all the visible regions of the sky. But the use of Dobsonian telescopes is limited to visual observation. It is hardly possible to motorize them or make them work

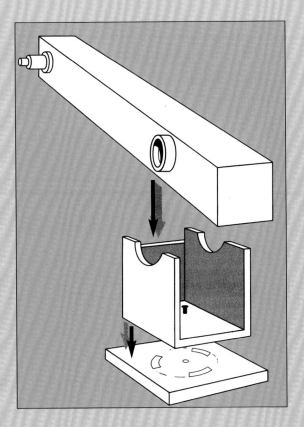

Diagram showing assembly of a Dobsonian telescope
This very simple system allows access to all regions of the sky.

automatically, which precludes the taking of photographs.

Some models have now been commercialized by the big telescope manufacturers. You can find instruments 8 inches (200 mm) in diameter, completely fitted with accessories (eyepieces, telescopic sight), for under $600.

It should be noted, however, that because of their primitive nature, Dobsonians can only be handled by observers who are very familiar with the sky and are used to finding faint objects among the brighter stars. Telescopic sights of the "starpointer" variety are precious aids that considerably simplify the task and make observation so much easier. All in all, the Dobsonian is a great advance, but you do need to know what you are looking for when using it.

What to observe?

Everything, of course, but especially faint objects: a Dobsonian telescope is an excellent instrument for the Messier Marathon (see p. 50)! However, since the observer has to follow the object manually, it is difficult to make a planetary drawing or take measurements of double stars. With plenty of practice, you can start to estimate the magnitude of variable stars, asteroids or comets. Remember that Dobsonians are only for viewing and do not allow you to take photographs.

The Moon
A 12-inch (300 mm) Dobsonian permits clear observation of the lunar surface: here, the Theophilus crater is seen.

M57
A 12-inch (300 mm) telescope enables you to observe faint objects, like this nebula, very well, as can be seen here.

Introduction to astronomy

Discovering the Big Bang

Astronomers and physicists have always been confronted by two big questions: where does our Universe extend to, and how was it born?

For the first philosophers, the Universe was limited to the Earth, which was enclosed in a frame encrusted with stars, the fixed sphere. Then, it was extended to the solar system, and the Earth found itself a simple planet among others, turning around the Sun. In the Renaissance, it was suggested that the stars were themselves other suns, increasing the size of the cosmos even further. Until the beginning of the 20th century, it was thought that the Universe was limited to our galaxy, the Milky Way, consisting of some 200 billion stars, surrounding us in a vast disk tens of thousands of light years wide. Even Albert Einstein still considered the Universe to be static and restricted in volume. But the setting up of the first giant telescope – the Mount Wilson observatory, which had a mirror with an 8 foot (2.5 m) opening – brought a real revolution in its train. In 1924, the astronomer Edwin Hubble noted that the renowned Andromeda Nebula (a milky blotch in the constellation of the same name) was not a gas cloud belonging to the Milky Way, but another galaxy in the background, itself teeming with stars! As other spots were also revealed to be sister galaxies to our own, the Universe saw its dimensions explode, with distances henceforth expressed in millions of light years.

A crowd of galaxies
The Abell 2218 cluster, with its gravitational arcs, gives a very good idea of the depth of the Universe.

An expanding Universe

While analyzing the light from these new galaxies, the astronomer Vesto Slipher noticed a strange phenomenon: their radiance was redder than that of the Milky Way. This could be explained if the galaxies were moving away from us: due to the Doppler effect, their radiant frequencies would shift from blue toward red, just as a siren moving away into the distance shifts from a high-pitched to a deeper sound. The recession speeds of the different galaxies could even be calculated according to the extent to which they reddened. In compiling the results, Hubble ended up making a surprising observation: The further off the galaxies were, the faster they were moving away from us. Double the distance meant double the speed. There was only one explanation: The entire Universe was expanding, like the surface of a bubble getting bigger. You can check this by drawing marks on a balloon, then blowing it up: as the surface stretches out, the marks get more and more distant from one another; and the further one mark is from another, the quicker they separate.

The mystery of our beginnings

Such expansion has dizzying implications: if the film were "played backwards," the Universe would be compressed into a smaller and smaller space. From this we can deduce that at the beginning of time all the galaxies must have touched, merging into an ever hotter, ever denser, ever smaller ball. The physicist Georges Lemaître undertook some calculations and concluded that the entire Universe, in its beginning, fitted into a dot smaller than an atom. Apparently, a violent burst of expansion led to the Universe being created from that single dot and matter was spread in all directions: this is the theory ironically nicknamed the "Big Bang" by its detractors.

Other physicists, among them George Gamow, calculated the conditions (temperature, density and pressure) which prevailed during the first moments of the Universe. Thanks to their equations, they were able to go back to the first ten millionth of a billionth of a billionth of a billionth of a billionth of a second

Inflation of the Universe

The classic Big Bang theory comes up against some strange paradoxes. Thus, the mass of the Universe seems to be miraculously well "chosen": a bit denser, and the Universe would have fallen back on itself after only a few centuries, a little less dense, and it would have been "diluted" very quickly - too quickly to allow the formation of stars and galaxies. But it avoided these apparently very likely catastrophes - it appears to be literally made to measure.

From the 1980s onwards, physicists, particularly Alan Guth and Andreï Linde, found an explanation for this paradox while working on quantum physics equations. According to them, 10^{-39} seconds after its birth, the Universe discharged a colossal amount of energy - literally springing up from nothing - which phenomenally accelerated its expansion. This "inflation" saw space grow from a subatomic size to that of a grapefruit in a tiny fraction of a second (10^{-35} s). This over-inflated space is then thought to have resumed normal expansion, the speed it had acquired preventing it from falling back on itself. Moreover, small variations in density, inevitable on a microscopic scale, were expanded and preserved when this inflation occurred, and it was from these "small lumps" that future galaxies were to be born.

(10^{-43} s) of its existence. The Universe was then 10 million billion billion times smaller than an atom of hydrogen, but it contained phenomenal energy: the temperature was more than 1 million billion billion billion degrees Kelvin.

After one millionth of a second, the temperature fell to 10,000 billion degrees Kelvin, the energy having been converted into a staggering number (1,087) of quarks, which are the elementary particles of matter. As the temperature continued to drop, these quarks merged to form protons, neutrons and other electrons.

The Universe was hardly one second old when the protons and neutrons in turn merged to create atomic nuclei of deuterium (heavy hydrogen), helium and lithium. Then things calmed down: it was 300,000 years – the time for the temperature to drop beneath the threshold of 3,000 K – before the electrons combined with the nuclei and produced neutral atoms. Until then, the Universe remained opaque: bright photons could not pass through it because they came up against the free electrons. But as soon as these electrons were captured by the nuclei, the Universe became transparent to radiation.

Proof of the Big Bang

This Big Bang theory had the advantage of being verifiable: By the time 300,000 years had passed (and the temperature had fallen below 3,000 K), the whole Universe had become bathed in the fantastic radiation released into space. But since the Universe continued expanding for several billion years, the radiation saw its wavelength drawn out proportionately. Physicists calculated that "fossil" radiation must have continued in the form of a radio emission with a wavelength of several centimeters,

Clusters of galaxies
Spiral galaxies and giant elliptical galaxies populate this region of the sky, known as Coma Berenices (Berenice's Hair).

A vast expanse
The Hubble Space Telescope gives a vertiginous image of the deep sky, each one of these galaxies including millions of stars.

corresponding to an apparent temperature of only 3 K (−270°C or −454°F). In fact, this was detected in 1965 – but by accident. While testing a new radio aerial, the engineers Arno Penzias and Robert Wilson noticed a strange background noise, coming from all directions of the sky. This noise had a wavelength of several centimeters, and it was exhibiting a cosmic temperature of around 3 K. It was the imprint of the Big Bang.

What next?

By continuing its formidable expansion, the Universe has now reached a respectable age: 15 billion years. The question is whether it will continue its expansion indefinitely, and at a rapid rate – in this scenario, the Universe is said to be "open." Or perhaps it will slow down, and then, under the burden of gravity, turn back and "implode," destroying all the galaxies (the "big crunch") until it melts back into a single dot – if this is the case, the Universe is regarded as "closed."

A final possibility is for the momentum for the flight of the galaxies to be counterbalanced exactly by their gravitational pull on one another; in such circumstances, the expansion of the Universe will progressively slow down, but without going backwards. It can then be described as "flat," because calculations have shown that such a Universe would not have any curvature that was noticeable on a large scale; straight lines would remain straight lines, etc. To determine which of these three hypotheses is the right one, astronomers are attempting to observe the Universe's geometry and to estimate its total mass, since it is the Universe that dictates whether it is open or closed. If it is too weak, it cannot fight expansion, which will continue forging ahead; if it is too strong, it will win this titanic battle and collapse in on itself.

Is the Universe open or closed? It is now in a position to yield up its secrets (see following pages).

What is the structure of the Universe?

The Big Bang theory gives a convincing explanation of the birth and expansion of the Universe. But great enigmas still remain: How did matter collect in various places to form individual concentrations and create galaxies? What is the fate of the Universe? How can we measure its total mass and decide whether this is slowing down its expansion to the point of stopping it and making it fall in on itself? The observations of large telescopes and satellites are now giving us the first answers.

The structure of the Universe is the first problem. By counting the galaxies, astronomers have discovered that they are grouped in clusters, superclusters and great "walls," forming a complex pattern, with "inner walls" rich in galaxies and empty regions between them. We have evidence that this pattern is inherited from the Big Bang. The fossil radiation at 3 K, reflecting the state of the Universe 300,000 years after its birth, shows fluctuations (or anisotropy) of one hundred thousandth of a degree Kelvin, heralding the creation of these future pockets of matter.

The mystery of the missing mass

In the 1930s, the American Fritz Zwicky showed that the mass of visible stars was not sufficient to explain the movement of the galaxies, in particular their rotation and that of the clouds surrounding them. The galaxies behave as if they were 10 times greater in mass than their numbers of stars would lead us to believe. This surplus invisible matter, or "dark matter," is not only contained in the galaxies. If we are to go by the complex movements of clusters, it is also to be found throughout space.

This dark matter does not consist of gas clouds and dust; if this were the case, it would be obvious, because they would send out radio and infrared waves. The dark matter might consist rather of "failed" stars, which are cold and too

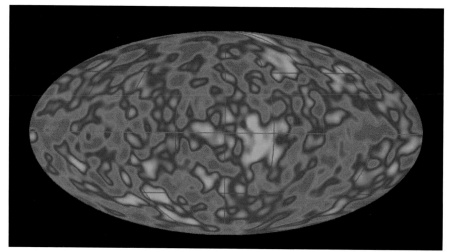

The Universe according to COBE
The COBE satellite has observed the fluctuations of cosmic background radiation throughout the Universe.

compact to be spotted, and which have been named "brown dwarfs." But the tracking of these hypothetical objects has not yielded any convincing results. Physicists are therefore wondering whether the Universe is crammed with invisible particles that might have a mass. Two candidates for this "exotic" matter have emerged from their calculations: "neutralinos" and "axions," which might represent in total six to seven times the mass of ordinary matter.

A mysterious force of repulsion

However, this is still not enough to explain the structure of the Universe, which has a "flat" geometry. This means that its force of gravity should provide a perfect counterbalance to the flight of the galaxies. Dark matter and exotic matter only represent 30% of the mass necessary to achieve this equilibrium. Physicists are therefore wondering if the deficit might be compensated for by a strange form of energy, which has been named "dark energy."

Recent observations confirm this hypothesis. Comparing the speeds of flight of the distant galaxies – which date from the early stages of the Universe – to those of the younger galaxies, their movement does indeed appear to speed up over time, under the influence of a mysterious force of repulsion.

The X-ray Universe The sources of X-rays in the Universe have been observed by the satellite XMM, the telescope of the European Space Agency.

In conclusion, the Universe was born more or less flat, balanced between its expansion and the slowing down effect due to its mass. But the repulsive energy gradually gained ground over gravity and accelerated expansion. The Universe is becoming ever more "open," dilute and infinite.

Look-out satellites

COBE
COBE (Cosmic Background Explorer), the satellite designed to observe cosmic radiation, was launched in 1989.

The creation, structure and fate of the Universe are scrutinized by astronomical satellites, which dissect the Big Bang's fossil radiation. In 1992, COBE drew up the first map of the early Universe, showing the irregularities of its fossil radiation in the form of temperature variations of a hundred thousandth of a degree Kelvin. Since then, balloon probes have mapped regions of space with even more precision. Great things are expected from COBE's successor, the MAP (Microwave Anistropy Probe) satellite, which was launched in June 2001. It is much more sensitive and measures temperature variations five times smaller than those measured by COBE. Similarly, its angular resolution (fineness of detail) is ten times greater. These capabilities will allow us to determine the mass of the Universe to within 5%, and thus to ascertain its geometry and its fate.

2001/03/29 09:36 UT

May

Eruptions
SOHO is permanently observing sunspots. This picture from March 29, 2001 reveals intense activity.

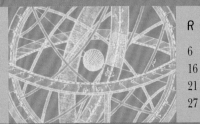

A month with your head up in the stars

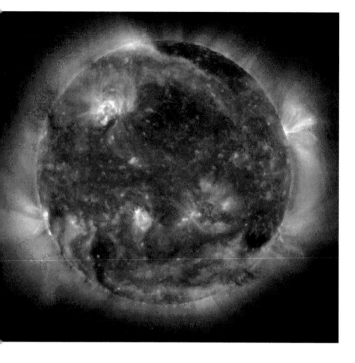

This month of May offers the wonderful sight of a total eclipse of the Moon, which can be observed from Europe under excellent conditions. The interplay of shadows and penumbra is accompanied by the occultation of a star, and demonstrates the Moon's speed of motion among the stars.

Tricolor Sun
A composite image of the Sun obtained from observations conducted by the SOHO satellite in three ultraviolet wavelengths.

Observations of the month

It is the constellation of Leo, together with Regulus, Virgo and Spica, which dominates the early part of these nights. Vega appears in the northeast, announcing the Summer Triangle, and Ursa Major is very high above our heads. Scorpius, with Antares, gradually appears in the southeast... Summer is approaching. On the western horizon, Gemini disappears at the start of the night, together with Saturn.

The planets
The Moon, **Venus**, **Mars** and **Saturn** will all be assembled for our enjoyment on the 21st

All times are given in Eastern Daylight Time (EDT). Calculations have been made for a latitude of 44° north and a longitude of 80° west.

TIMES OF RISING AND SETTING FOR THE SUN, MOON AND FIVE PLANETS VISIBLE WITH THE NAKED EYE														
Date	Sun		Moon		Mercury		Venus		Mars		Jupiter		Saturn	
	Rises	Sets	Rises	Sets	Rises	Sets	Rises	Sets	Rises	Sets	Rises	Sets	Rises	Sets
5/1	6:12 a.m.	8:20 p.m.	4:41 p.m.	4:47 a.m.	5:33 a.m.	6:28 p.m.	7:56 a.m.	12:08 a.m.	8:35 a.m.	12:13 a.m.	2:43 p.m.	4:05 a.m.	9:40 a.m.	12:59 a.m.
5/5	6:07 a.m.	8:25 p.m.	8:39 p.m.	6:24 a.m.	5:23 a.m.	6:16 p.m.	7:48 a.m.	12:02 a.m.	8:30 a.m.	12:09 a.m.	2:27 p.m.	3:49 a.m.	9:26 a.m.	12:45 a.m.
5/10	6:01 a.m.	8:31 p.m.	2:25 a.m.	11:22 a.m.	5:13 a.m.	6:11 p.m.	7:37 a.m.	11:50 p.m.	8:25 a.m.	12:03 a.m.	2:08 p.m.	3:30 a.m.	9:08 a.m.	12:27 a.m.
5/15	5:55 a.m.	8:37 p.m.	4:27 a.m.	5:15 p.m.	5:04 a.m.	6:14 p.m.	7:23 a.m.	11:33 p.m.	8:19 a.m.	11:57 p.m.	1:49 p.m.	3:10 a.m.	8:51 a.m.	12:10 a.m.
5/20	5:50 a.m.	8:42 p.m.	6:26 a.m.	9:34 p.m.	4:58 a.m.	6:24 p.m.	7:07 a.m.	11:11 p.m.	8:14 a.m.	11:50 p.m.	1:31 p.m.	2:51 a.m.	8:34 a.m.	11:52 p.m.
5/25	5:46 a.m.	8:47 p.m.	10:54 a.m.	1:37 a.m.	4:53 a.m.	6:40 p.m.	6:47 a.m.	10:43 p.m.	8:10 a.m.	11:43 p.m.	1:12 p.m.	2:36 a.m.	8:17 a.m.	11:35 p.m.
5/30	5:42 a.m.	8:52 p.m.	4:48 p.m.	3:31 a.m.	4:51 a.m.	7:09 p.m.	6:24 a.m.	10:09 p.m.	8:05 a.m.	11:36 p.m.	12:54 p.m.	2:13 a.m.	8:00 a.m.	11:17 p.m.

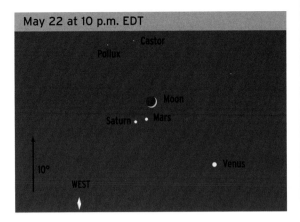

May 22 at 10 p.m. EDT

Castor
Pollux
Moon
Saturn · Mars
10°
Venus
WEST

and 22nd, in the early part of the evening. This will be the moment to get your bearings and, through observations carried out on subsequent days, measure the relative motions of these heavenly bodies, which our ancestors named the planets, or "moving stars." Look closely at how Venus seems to be moving nearer the Sun, ready to pass in front of it in June, an event that has not occurred for 120 years. On the evening of the 22nd, look out for a very attractive grouping on the western horizon: Saturn, Mars and the Moon are above Venus and below the Twins, Castor and Pollux. The next setting of the Dioscuri heralds the rise of Cygnus, the Swan, with the promise of

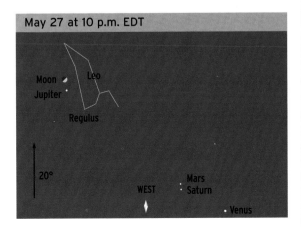

May 27 at 10 p.m. EDT

Moon Leo
Jupiter
Regulus

20°
Mars
WEST Saturn
Venus

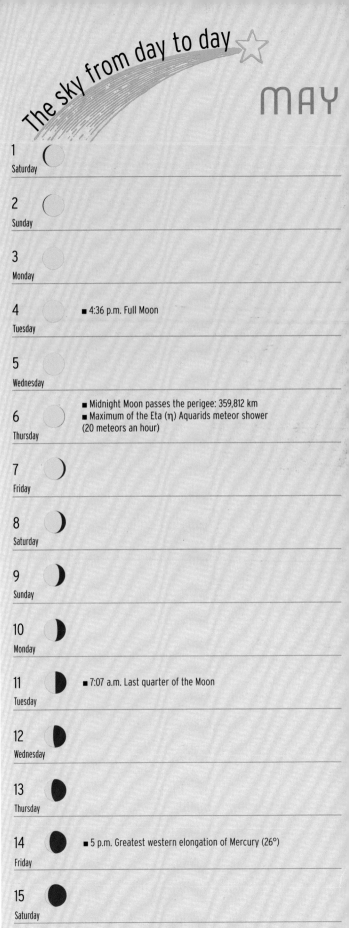

1 Saturday

2 Sunday

3 Monday

4 Tuesday
■ 4:36 p.m. Full Moon

5 Wednesday

6 Thursday
■ Midnight Moon passes the perigee: 359,812 km
■ Maximum of the Eta (η) Aquarids meteor shower (20 meteors an hour)

7 Friday

8 Saturday

9 Sunday

10 Monday

11 Tuesday
■ 7:07 a.m. Last quarter of the Moon

12 Wednesday

13 Thursday

14 Friday
■ 5 p.m. Greatest western elongation of Mercury (26°)

15 Saturday

Aurora australis
These southern lights were photographed in New Zealand in November 1998.

lovely summer evenings. On the 23rd, to close the performance, the Moon passes 2° away from Pollux; and on the 26th, at 8 p.m., it is 4° away from Regulus, the principal star in the constellation of Leo, and 10° away from Jupiter. By the 27th, **Jupiter** and Regulus will have been overtaken by the Moon, which is getting larger, and will continue its course toward the constellation of Virgo, to reach it by the end of the month.

The polar aurorae

The polar aurorae are of exceptional beauty, and not to be missed. In Classical times, they were thought to be caused by the spirits of the dead abandoning themselves to dancing and games. These phenomena occur simultaneously in the north (aurora borealis or northern lights) and in the south (aurora australis or southern lights).

Until the end of the 19th century, it was thought they were luminous reflections caused by the polar ice or by ice crystals in the Earth's atmosphere. It was only in the 20th century that people came to understand that the Sun played a primary role in these phenomena. Our

planet possesses a magnetic field, whose orientation is at present determined by the polar axis. Electrically charged particles are subjected to this magnetic field and channeled toward the poles. When there is a solar eruption, a large number of particles, mainly electrons and protons, leave the Sun at speeds that can attain several hundred kilometers per second. In a few dozen hours, these particles arrive in the vicinity of the Earth and can be trapped by our planet's magnetic field. The interaction between these particles and the atoms in the Earth's atmosphere can be violent. The atmosphere then lights up like an electric tube and displays colors characteristic of the excited atoms that are present (mainly nitrogen and oxygen), which radiate in the green and the red. The phenomenon occurs at very high altitude, between 100 and 300 kilometers from the ground, making it extraordinarily impressive. When solar activity is intense, the electrically charged solar particles can reach relatively low latitudes. Polar aurorae can be predicted by the observation of solar eruptions, and there are Internet sites that give these forecasts (www.spaceweather.com or www.dcs.lancs.ac.uk/iono/aurorawatch). Be sure to consult these if you don't want to miss the sight.

Gemini center stage

Castor and Pollux, the two main stars in the constellation of Gemini, are classified 23rd and 17th respectively among the most luminous stars in the sky. The pair are separated by only 4.5°. Castor is the more northerly and the less luminous of the two. According to legend, Castor and Pollux were the sons of Leda, but had different fathers, Zeus himself being the father of Pollux. It was in the guise of a swan that Zeus approached the beautiful Leda, and this is why the constellation of Cygnus only appears when the Twins disappear over the western horizon. Castor is about 51 light years from the Sun, whereas Pollux is much nearer, at 35 light years. Castor is a multiple star: six components are known, the two main ones being visible with a small instrument (separation 3.9") and these revolve around one another over a 500-year period. In fact, each of these two stars is itself a double. The first is composed of two practically identical stars revolving around each other in a little over nine days, and the second is made up of two stars which revolve around each other in a little under three

M35
This beautiful open cluster, observable with a small instrument, is in the constellation of Gemini about 2,200 light years away from the Sun.

16 Sunday	■ The Moon and Mercury move closer together. Visible at the end of the night (minimum 2.6° at 7 p.m.)
17 Monday	
18 Tuesday	
19 Wednesday	■ 12:54 a.m. New Moon
20 Thursday	
21 Friday	■ 8 a.m. Moon passes the apogee: 406,262 km ■ The Moon and Venus move closer together. Visible in the early part of the night (minimum 0.3° at 8 a.m.)
22 Saturday	■ The Moon and Mars move closer together. Visible in the early part of the night (minimum 3.2° at noon)
23 Sunday	
24 Monday	
25 Tuesday	
26 Wednesday	
27 Thursday	■ 4:02 a.m. First quarter of the Moon ■ The Moon and Jupiter move closer together. Visible in the evening (minimum 3.7° at 8 a.m.)
28 Friday	
29 Saturday	
30 Sunday	
31 Monday	

days. The system possesses a distant companion, Castor C, which orbits the four "central" stars in about 10,000 years. Castor C is also a double, its two components being "red dwarfs," which revolve around each other in less than 20 hours. Pollux is quite a large star with a diameter about five times bigger than that of the Sun. It is 35 times more luminous than the Sun and its surface temperature is of the order of 4,500 K. The "Twins" are clearly much more complex than the Ancients believed when they first identified them centuries ago!

Shooting stars

The **Eta** (η) **Aquarids** are caused by debris from Halley's Comet. This meteor shower has been known since Classical times, for observations have been found dating from 401 BC. The maximum on the 6th is not very striking, because appearances are spaced out between April 21 and May 12, but a few bolides may occur toward the end of this time. Halley's Comet is also the cause of the **Orionids** meteor shower in October. So this comet, which is only visible every 76 years, indirectly reminds us of its existence twice a year thanks to shooting stars.

A region of the sky to explore: the constellation of Hercules

Like Orion, Hercules (the Roman equivalent of the Greek Herakles) is a very large constellation and represents a hero of more than ordinary strength. Hercules played such an important role in mythology that he inspired astronomers when they drew up charts of the sky and named the stars and constellations. It was Hercules who, placed by Mercury on the breast of the sleeping Juno, hoped to gain immortality by drinking the milk of the goddess. When the latter suddenly awoke, the flow of milk spurted heavenwards, causing the Milky Way to appear in the sky. What is more, the constellations of Leo and Hydra represent two of the labors of Hercules: his struggle with the Nemean Lion and the Hydra of Lerna.

Rasalgethi (or **Alpha** (α) **Herculis**), Arabic for "the head of the person kneeling," is one of the brightest among the irregular variables, going from a magnitude of 3.1 to 3.9. This variability was established by William Herschel in 1795. Its average period is 90 days, and the star is more than 430 light years away, its diameter being at least 400 times greater than that of the Sun. It is also a very beautiful double and, when observed with a small 2.5-inch (60 mm) telescope, presents an attractive color contrast going from red to blue-green. Its companion, situated at 4.6", has a magnitude of 5.4. The period of the system is about 4,600 years.

Beta (β) **Herculis** is an excellent example of an optical double. Its companion, with a magnitude of 8, is 8.9" from Beta, which has a magnitude of 3.14. The two stars move in different directions, but the apparent angular distance between them is at present diminishing. Once again, there is a lovely color contrast of green and mauve, visible with a 2.5-inch (60 mm) instrument.

THE SKY IN MAY

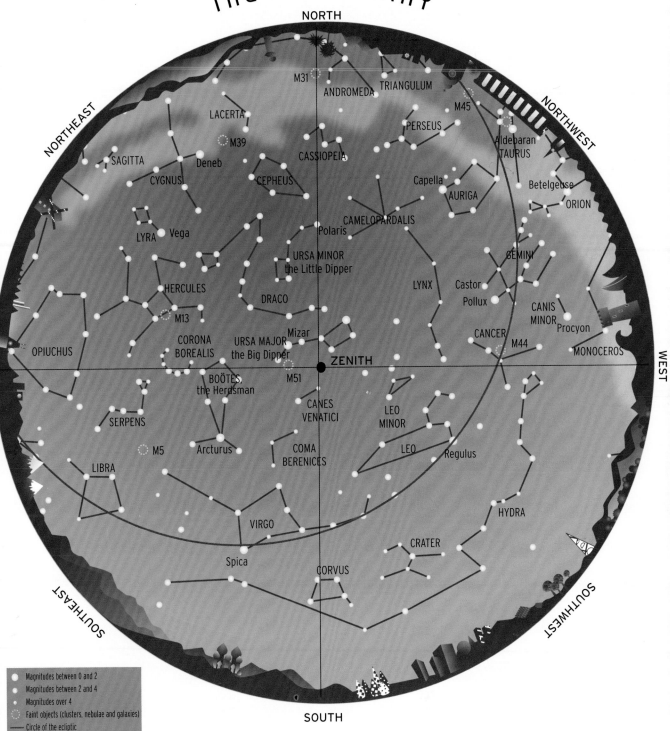

NORTH

NORTHEAST

M31 · ANDROMEDA · TRIANGULUM

LACERTA · M45 · NORTHWEST

SAGITTA · M39 · PERSEUS

CYGNUS · Deneb · CASSIOPEIA · Aldebaran · TAURUS

CEPHEUS · Capella · Betelgeuse

LYRA · Vega · CAMELOPARDALIS · AURIGA · ORION

Polaris

URSA MINOR · GEMINI

the Little Dipper · LYNX · Castor · CANIS

HERCULES · DRACO · Pollux · MINOR

M13 · Mizar · Procyon

CORONA · URSA MAJOR · CANCER · M44

OPIUCHUS · BOREALIS · the Big Dipper · MONOCEROS

WEST

ZENITH

BOÖTES, · M51

the Herdsman · CANES · LEO

SERPENS · VENATICI · MINOR

M5 · Arcturus · COMA · LEO

BERENICES · Regulus

LIBRA

VIRGO · HYDRA

CRATER

Spica · CORVUS

SOUTHEAST

SOUTHWEST

SOUTH

○ Magnitudes between 0 and 2
○ Magnitudes between 2 and 4
· Magnitudes over 4
⊙ Faint objects (clusters, nebulae and galaxies)
— Circle of the ecliptic

■ **How to use this chart**
Hold the chart above your head, matching the word SOUTH that appears at the edge of it with the geographical south of the place you are observing from. Use a compass to help you do this.

■ **This chart shows the sky that is visible at a latitude of 45°**
If you are further north or further south, Polaris will be higher or lower in the sky.

■ **Chart of the sky visible at 11 p.m. EDT** at the beginning of the month; at 10 p.m. EDT in the middle of the month; at 9 p.m. EDT at the end of the month.

Two clusters in the constellation

M13 is one of the best-known objects in the sky and the most beautiful example of a globular cluster (that of Hercules) visible in the northern celestial hemisphere. It was first mentioned by Edmund Halley in 1715. Charles Messier cataloged it in 1764 and described it as "a nebulosity containing no stars," which shows that he had only poor-quality instruments at his disposal. M13 is visible to the naked eye on a good night without a Moon. This large cluster is to be found a third of the way along the line from Eta (η) to Zeta (ζ). It is easy to make out with binoculars, but more detail is revealed when you use a more powerful instrument. Here is an easy way to judge the quality of your equipment: through a small telescope, you can make out a bright circular nebula of about 10" in diameter; with a 5-inch (120 mm) telescope, the cluster turns out to be much more extensive and begins to resolve into stars; above 8 inches (200 mm), resolution is easy, but it is with the much larger instruments of above 12 inches (300 mm) that the spectacle is most striking. With these, you get a real impression of relief and you feel you are discovering the third dimension of the sky in all its depth. On photographs, M13 extends further still, and the sensation of depth is accentuated by the presence of distant galaxies visible on its periphery. The diameter of M13 is about 165 light years, and its distance from Earth is 25,100

The constellation of Hercules

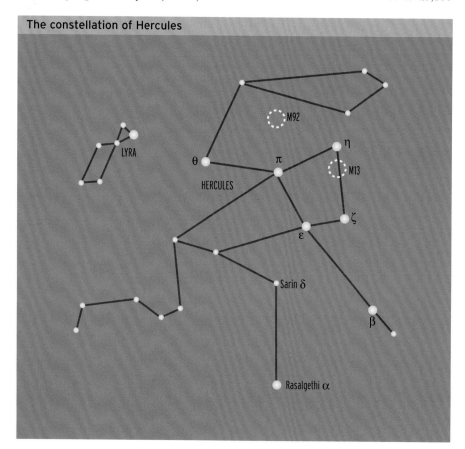

M13
This globular cluster in the constellation of Hercules is one of the wonders of the sky in the northern hemisphere.

light years. It contains more than a million stars, the central concentration being about 500 times denser than that near the Sun. M13's precise age is uncertain, but there is no doubt that it is very old: the American astronomer Alton Arp puts it at 14 billion years. But M13 contains a young blue-colored star, Barnard 29, which seems to have been captured by the enormous swarm of old stars.

In 1974, M13 was chosen to receive a message addressed to a possible extraterrestrial intelligence. This type of communication is transmitted by the giant radio telescope at Arecibo, in Puerto Rico. Once the message sent to M13 is received, and if the recipients

respond, we should not expect any return mail within 50,000 years!

M92, discovered by Johan Bode in December 1777, then cataloged by Messier in March 1781, is a very beautiful globular cluster which suffers from being close to its celebrated neighbor, M13. It is nevertheless visible to the naked eye against a dark sky without a Moon, and is easy to see with binoculars, being hardly any further away than M13, with a distance of 26,700 light years. The two clusters are in fact separated by 9°, with M92 to the northwest of M13. With a diameter of 85 light years, it contains several hundred thousand stars. Who knows, perhaps some of these may have planetary systems.

Practical astronomy

How to observe the Sun

The Sun is an exceptional star and the observation techniques used in other areas of astronomy cannot be applied here. From Classical times, it was understood that, if you wanted to examine the course of the Sun, you had to look at a shadow. In order to look straight at the Sun, special precautions must be taken; its light is so powerful that it is essential for this protection to be effective.

With the naked eye

It is dangerous to observe the Sun with the naked eye, because our retina is not innervated: in a few seconds, the image can burn several cells without our feeling any pain. The damage is irreparable and can manifest itself through symptoms such as permanent black dots on the field of vision. The only possible way of observing the Sun – but an important way – is to observe its movement. For thousands of years, humans have placed markers on the Earth allowing them to study the course of the Sun. The simplest is the "gnomon," a rod planted vertically in the ground. The shadow changes in size and direction throughout the day, but the direction of the shortest shadow is always the same: it is the meridian. Reading this shadow is the same as interpreting the shadows that characterize each day – because the length of the meridian shadow changes from day to day, between the two solstices. With a little perseverance and a regard for accuracy, you will be able to measure out the meridian and produce a natural calendar. Thanks to the gnomon and the shadow, it is possible to determine the date on the meridian line naturally.

With optical instruments

Observing the Sun with telescopes demands particular precautions, because the amount of energy gathered is even greater than with the naked eye. As proof of this, the image of the Sun obtained through a small magnifying glass can set fire to a piece of wood within a few seconds.

As soon as the first optical instruments appeared, projection techniques were used. The principle is very simple: thanks to a traditional lens system, the image of the Sun is projected onto another medium, and this is the image that is looked at. The energy is therefore spread over a larger surface and no longer presents the dangers of being concentrated in the lens of an instrument. Whether with binoculars, or a small or

Projection
The Sun's image is projected through a telescope onto a screen, where it is possible to observe sunspots with the naked eye.

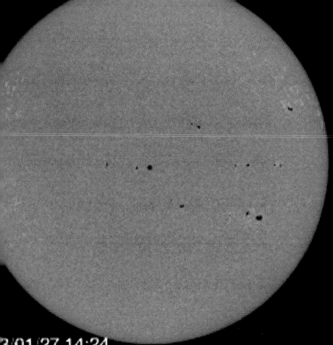

3/01/27 14:24

Sunspots
The Sun is observed here by SOHO, on January 27, 2003.

What should you observe?

Observing the Sun essentially means studying sunspots. It is interesting and important to measure the appearance and evolution of these ephemeral curiosities. By doing this, you can measure the rotation of the Sun and predict when the polar aurorae will appear: some of these are visible as far south as a latitude of 35°. The appearance of a sunspot at the center of the disk is in fact followed, within two or three days, by a massive arrival of electrical particles into the Earth's atmosphere, causing a polar aurora.

Enthusiasts can observe the Sun through interferometric filters; these are usually filters known as Hs, which isolate the distinctive hydrogen line and enable you to observe solar prominences, in the form of filaments of matter attached to the edge of the disk. Investigating these prominences and the lower corona is made easier by the "coronograph": you place a metal disk, the exact size of the Sun's image, in the lens of the instrument to cause an artificial eclipse by optical means. The image of the solar disk is hidden and you see only the lower corona, exactly as in a total eclipse. You are in a way putting an artificial Moon in the field of vision. Instruments of this type (about $200) can be adapted to the lenses of amateur telescopes. It is possible to observe the Sun on the ESA (European Space Agency) website, which has been publishing pictures obtained by the SOHO probe since December 1995 (www.esa.int).

larger telescope, this image, projected onto a screen, enables you to admire sunspots.

Especially to be avoided are eyepiece filters, which can be placed behind the light collector, near the instrument's lens, where the heat concentration is greatest. They often come with commercial telescopes, and are extremely dangerous for they are liable to break due to the effects of the temperature; so during observation, they can let an enormous amount of energy into the eye suddenly, which can cause irreversible lesions. You should use lens filters – sun filters placed in front of the lens. These come in two sorts. There are filters that are strips of glass, mostly colored, and polished so as not to affect image quality. These are fragile and quite expensive: about $50 for a diameter of 4 inches (100 mm). Then there are filters made from sheets of aluminized Mylar, which are fastened over the lens of the instrument. Mylar is not very thick, so image quality is not greatly affected and, although fragile, these filters are relatively cheap: a 15 centimeter square sheet of Mylar costs about $40.

Solar eruption observed by SOHO

Introduction to astronomy

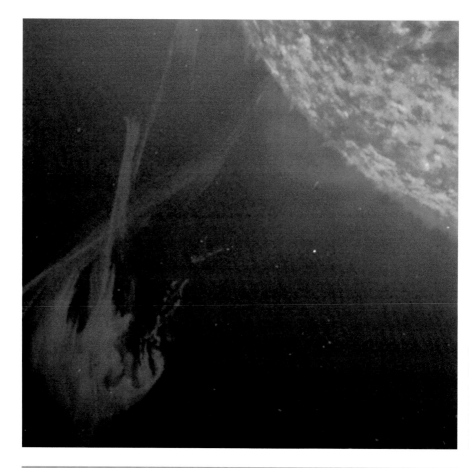

Solar prominence
This jet of incandescent matter, observed by SOHO, measures 500,000 km.

Discovering the Sun

The Sun is a rather small star, with a diameter of 1.39 million kilometers, that is 109 times that of the Earth, and a mass equivalent to 300,000 times that of our planet. Its size and the distance separating it from us (150 million kilometers) were finally discovered in the 17th century. Its light takes 500 seconds to reach us.

The Sun is 4.56 billion years old. It takes about one month to revolve completely: 25 days at the equator, and 33 days near the poles. This difference shows that it is fluid and loses its shape in the way a whirlwind does. By observing the spots scattered over its surface it has proved possible to measure its rotation speed. We can tell its

temperature from its color: on the visible surface, this is 5,700 K. The energy it sends out, providing us with light and heat, comes from what is a veritable nuclear boiler at the Sun's center, where hydrogen is transformed into helium by fusion reactions at a heat of 15 million degrees Kelvin. This "active" center is concentrated in a core whose radius is more or less equal to a quarter of the Sun's radius, or 175,000 kilometers.

Like all stars, the Sun is chiefly made up of hydrogen (92.1%), helium (7.8%) and traces of other elements (less than 0.1%). Like them, it is not eternal. It will disappear, absorbing or destroying its cortege of planets, and returning the matter of the solar system to the

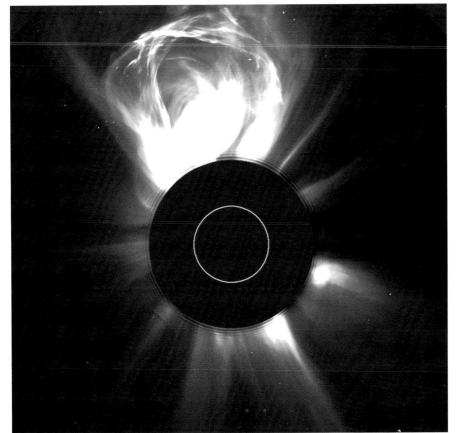

The lower solar corona
This region is observed by SOHO with the aid of a coronograph, which eclipses the solar disk.

interstellar environment, in which new stars can be formed.

From center to surface

From the core, energy is projected out into a vast intermediate radiative zone extending over nearly 90% of the Sun's radius, and rising up to 600,000 kilometers. Beyond it is the convective zone, where energy is carried toward the surface owing to considerable movements of matter. The outer layers are very thin, with the photosphere (about 500 kilometers wide) providing the transition to the chromosphere (about 2,500 kilometers). The photosphere is the Sun's visible surface. Beyond it is the corona that extends into the solar system. The spots are gigantic gas bubbles, which rise and burst on the Sun's surface. They are 1,000 K to 2,000 K lower in temperature than the surrounding matter and are only visible for a few days. They generally appear in groups and are the focus of intense magnetic activity. Their number, measured at any given moment, is an indicator of solar activity, which occurs in 11-year cycles. Above these spots are large protuberances, which are only visible from Earth at the edge of the solar disk. Some can rise to several hundreds of thousands of kilometers. The matter that leaves the Sun in these veritable cataclysms makes up what is known as "the solar wind," a current of electrically charged particles moving away from the Sun at speeds of

The next accessible total eclipse of the Sun will take place on March 29, 2006. It will be visible within a band going from Natal (Brazil) to Sivas (Turkey), and passing through Accra (Ghana), Gusau (Nigeria), Maradi (Niger), near to Tobruk (Libya) and Kayseri (Turkey). The eclipse will be partial in Rome, Paris and Moscow.

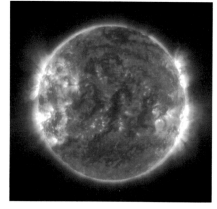

Solar activity
The Sun is observed here by SOHO at a wavelength of 171 nanometers, which corresponds to that of iron that has been ionized nine times.

several hundred kilometers per second. Some of these particles reach the Earth in a few days and, trapped by the Earth's magnetic field, give rise to the polar aurorae. Contrary to widespread belief, these aurorae are not only visible at high latitudes. When solar activity is intense, it is possible to see them at medium latitudes. Anyone can then take photographs of these remarkable manifestations, but the pictures obtained to do not entirely do justice to their size and beauty. Information about when aurorae may next be visible in your area can be obtained from the Internet (see p.37).

The corona

Around the Sun is the corona. The distinctive feature of this thin atmosphere is that it is at a much higher temperature than that of the surface, several million degrees Kelvin in fact. The corona extends a long way into the solar system, well beyond the Earth. But the remarkable impression it creates will only be available to those able to see total eclipses whenever they occur in the future, since these are the only circumstances under which it can be seen from Earth.

Eclipses occur relatively frequently, but total eclipses are rarely accessible to amateurs as they are often observable only in remote parts of the world – and even then, the weather conditions have to be right!

The workings of the Sun

Many theories were advanced before the way the Sun works could be explained satisfactorily. In the 19th century, people finally understood that the Sun was a star like others. Spectroscopy showed that the chemical composition of the stars and the Sun was similar, and, as soon as the distances of the nearest stars became known, around 1840, it became obvious that the difference in appearance between the stars and the Sun was only a question of distance. In 1862, Lord Kelvin developed the theory that the Sun might draw its energy from its own gravitational contraction. As the laws of gravity already allowed calculation of the time taken by a mass to contract, he predicted that the duration of the Sun's life could be about a hundred million years, unless new heat sources, at present unknown, were smoldering away in "the

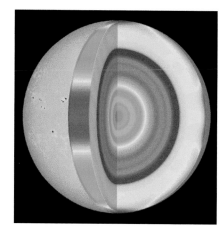

A cross section of the Sun
From the Sun's core to its surface, energy is transmitted through radiation and convection, the temperature going from about 13,000,000 K to 6,000 K.

The solar corona
It is only during a total eclipse, like this one in 1998, that we can get a direct view of the solar corona.

great melting pot of creation," as he described it. Then, at the beginning of the 20th century, with the discovery of radioactivity, came proof that certain terrestrial rocks were several billion years old, that is to say more than the age of the solar system imagined by Kelvin. It therefore became essential to find these "new heat sources." It was only in the 1920s, following the work of Einstein, that the thermonuclear hypothesis emerged and our understanding of the Universe was fundamentally altered.

The Sun radiates its energy thanks to nuclear fusion, which transforms four nuclei of hydrogen atoms into a nucleus of helium and into energy. The energy at present radiated corresponds to the consumption of four million tonnes of hydrogen per second. In reality, each second, 500 million metric tons of hydrogen are transformed into 496 million metric tons of helium, the difference being the energy radiated, in accordance with the principle of equivalence between mass and energy. The Sun will continue to radiate in this

way for about another five billion years. The end of its evolution will then cause the matter that has resided in it for 10 billion years to be ejected, resulting in the appearance of a planetary nebula. This will be centered on a white dwarf, the very hot skeleton of the star, whose size will be approximately the same as that of the Earth. In the constellation of Lyra, you can observe a magnificent example of a planetary nebula. This shows you what it will be like five billion years from now in the region occupied at present by our solar system.

The Sun in figures	
Average radius	6,955,080 km
Mass	$1,989 \times 10^{30}$ kg
Average density	1.409
Surface gravity	274 m/s^{-2}
Escape velocity	617.7 km/s
Surface temperature	5,777 K
Core temperature	15.7×10^6 K
Core density	151
Veocity in relation to neighboring stars	19.7 km/s
Age	4.6 billion years
Life expectancy	5 billion years

The Sun under scrutiny

SOHO
The Solar Heliospheric Observatory, SOHO, the European Space Agency probe, has been observing the Sun since 1995

From the beginnings of the space age, the Sun has been the target of exploratory missions with two complementary objectives. The first is to study how a star functions. The Sun is in fact an extraordinary laboratory where it is possible to study the mechanisms of nuclear fusion, the way matter behaves at very high temperatures, and the transfer of radiation. The other objective is to study interactions between the Sun and the Earth. Our planet, being near to the Sun, is bathed in its radiation and in the flux of particles escaping from it. A large part of the radiation and most of these particles never reach the ground thanks to the Earth's natural protective mechanisms. There are two kinds of protective mechanism: on the one hand there are the atmospheric filter and electromagnetic shield; on the other, the atmosphere and magnetosphere. Study of the magnetosphere and the upper atmosphere can only be carried out from space, where there is the possibility of observation on wavelengths that have been blocked by the Earth's atmosphere, in particular the ultraviolet.

Observation from space

After promising results from experiments conducted on board balloon probes, the first space observations have opened the way to a vast amount of knowledge. The very first American artificial satellite, Explorer 1, launched on January 31, 1958, revealed the presence of radiation belts surrounding the Earth: the Van Allen belts. It was the start of an exploration into the relationship between Sun and Earth.

The big missions

Since then, all the space agencies have carried out experiments involving the Sun. The joint ESA–NASA project, Ulysses, launched on October 6, 1990, explored the Sun's polar regions. The first flights over the poles were carried out in 1994 and 2000 for the south pole, and 1995 and 2001 for the north pole. Ulysses has considerably increased our knowledge of the solar wind, discovering in particular that its speed at high latitudes is much greater than previously thought.

The solar corona, which is almost unobservable from Earth except during total eclipses, has also been a target for important space missions. Coronas I, a Russian-Ukrainian vehicle, was launched on March 2, 1993, for ultraviolet and X-ray observation of the solar ionosphere and magnetosphere.

The most important missions at present being carried out involve studies of the Sun-Earth environment, and the relationship between the two. Cluster II is a group of four identical satellites: Rumba, Salsa, Tango and Samba departed on July 15 and August 9, 2000, with the help of Russian launchers after the first experimental launch of the Ariane V rocket failed. This quartet is exploring the solar wind and magnetosphere in order to produce the first three-dimensional representation of the Sun-Earth relationship.

Generis, which left on August 8, 2001, was put into orbit around L1, a gravitational equilibrium point between Sun and Earth, its job being to collect particles of solar wind. These samples will be brought back to Earth at the end of 2003.

SOHO (Solar and Heliospheric Observatory), launched on December 2, 1995, has been observing the Sun since the very beginning of 1996. Thanks to its extraordinary stability, it has enabled solar and high-energy physics to make great progress, thus opening the way for new experiments in space.

HESSI (High Energy Solar Spectroscopic Imager) was projected into space on June 21, 2000, for the X-ray observation of the acceleration of particles in solar eruptions, and it is planned for this to operate until 2004.

Finally, between 2008 and 2015, Solar Orbiter, a very ambitious ESA program, will be responsible for studying the Sun at close quarters, while collecting samples of solar wind particles.

The Sun has therefore been examined at all possible wavelengths for a number of years; an extraordinary invisible

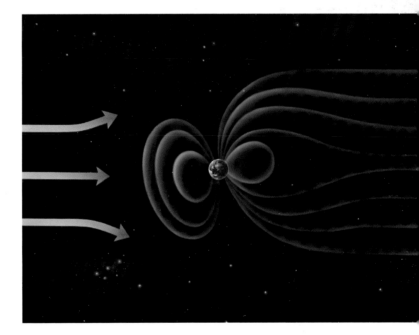

armada is tracking it above our heads without our being aware of it. Thanks to these programs, the Sun's mechanisms for producing and transferring energy are being better understood, as are the protective mechanisms which shield our planet from dangerous radiation – enabling it to continue to support life as we know it.

Our shield
The magnetosphere protects the Earth, guiding electrical particles from the Sun toward the polar regions.

June

NGC1288
This spiral galaxy, situated 300 million light years away, is moving away from us at a speed of 4,500 km/s.

RENDEZVOUS IN THE SKY

A month with your head up in the stars

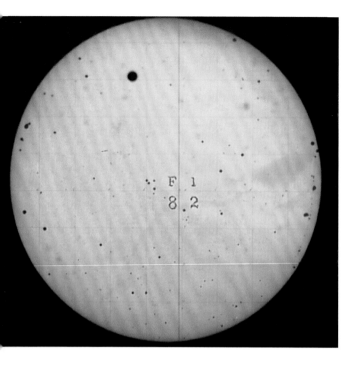

The event of the month is Venus's transit in front of the Sun, which you can observe on the morning of the 8th. The nights are becoming shorter, but reveal some very beautiful sights when the Moon and planets meet. It is also an ideal time to explore the zodiacal constellations of the South: Scorpius and Ophiuchus, the Serpent Bearer.

Transit of Venus
This photo was taken in Wellington, in South Africa, on the occasion of Venus's last transit in front of the Sun, on December 6, 1882.

Observations of the month

To observe the Moon rising is a must on the 2nd: at 9 p.m. it is 1.5° from Antares and M4, the beautiful globular cluster in the constellation of Scorpius. It is impressive to see it moving closer to the red star just with the naked eye, but through binoculars or a small telescope, the passage of the Moon, now almost full, is quite splendid.

The summer solstice is well worth observing too: on the 21st, the Sun rises as far as it can in the northeast. Look at the direction in which it rises and remember it so as to compare it with where it rises at the winter solstice, in six months' time.

All times are given in Eastern Daylight Time (EDT). Calculations have been made for a latitude of 44° north and a longitude of 80° west.

TIMES OF RISING AND SETTING FOR THE SUN, MOON AND FIVE PLANETS VISIBLE WITH THE NAKED EYE														
Date	Sun		Moon		Mercury		Venus		Mars		Jupiter		Saturn	
	Rises	Sets	Rises	Sets	Rises	Sets	Rises	Sets	Rises	Sets	Rises	Sets	Rises	Sets
6/1	5:41 a.m.	8:53 p.m.	7:31 p.m.	4:19 a.m.	4:51 a.m.	7:14 p.m.	6:15 a.m.	9:55 p.m.	8:04 a.m.	11:33 p.m.	12:47 p.m.	2:05 a.m.	7:53 a.m.	11:10 p.m.
6/5	5:39 a.m.	8:56 p.m.	–	7:44 a.m.	4:54 a.m.	7:38 p.m.	5:55 a.m.	9:24 p.m.	8:01 a.m.	11:26 p.m.	12:33 p.m.	1:50 a.m.	7:40 a.m.	10:56 p.m.
6/10	5:39 a.m.	8:59 p.m.	2:14 a.m.	2:01 p.m.	5:03 a.m.	8:06 p.m.	5:29 a.m.	8:44 p.m.	7:57 a.m.	11:18 p.m.	12:16 p.m.	1:31 a.m.	7:23 a.m.	10:39 p.m.
6/15	5:38 a.m.	9:02 p.m.	3:57 a.m.	7:28 p.m.	5:18 a.m.	8:42 p.m.	5:05 a.m.	8:06 p.m.	7:53 a.m.	11:09 p.m.	12:00 p.m.	1:13 a.m.	7:06 a.m.	10:22 p.m.
6/20	5:39 a.m.	9:04 p.m.	7:43 a.m.	11:40 p.m.	5:40 a.m.	9:16 p.m.	4:42 a.m.	8:23 p.m.	7:50 a.m.	10:59 p.m.	11:43 a.m.	12:54 a.m.	6:49 a.m.	10:05 p.m.
6/25	5:40 a.m.	9:04 p.m.	1:16 p.m.	1:15 a.m.	6:08 a.m.	9:44 p.m.	4:22 a.m.	6:51 p.m.	7:47 a.m.	10:50 p.m.	11:27 a.m.	12:36 a.m.	6:32 a.m.	9:47 p.m.
6/30	5:42 a.m.	9:04 p.m.	7:47 p.m.	3:23 a.m.	6:38 a.m.	10:03 p.m.	4:03 a.m.	6:32 p.m.	7:44 a.m.	10:40 p.m.	11:11 a.m.	12:18 a.m.	6:16 a.m.	9:30 p.m.

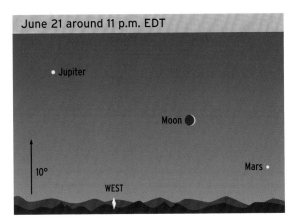

June 21 around 11 p.m. EDT

Jupiter

Moon

Mars

10°

WEST

Four days after the solstice, the Moon will be in its first quarter. Thus it will be where the Sun was three months earlier, at the time of the spring equinox. At the very beginning of the evening, the presence in the sky of **Jupiter**, **Mars** and **Saturn** enables us to see the route followed by the Sun since the equinox.

On the 23rd, at 7 p.m., the Moon, very near to Jupiter (4°), passes in front of M96, to find itself right next to M95 and M105. You should therefore explore the immediate vicinity of the Moon with binoculars or a spyglass, as soon as the Sun goes down, to try to see the galaxies appear: M95 is a spiral with a magnitude of 9.7, M96 is more luminous with a magnitude of 9.2, and M105 is an elliptical galaxy with a magnitude of 9.3. The three galaxies are

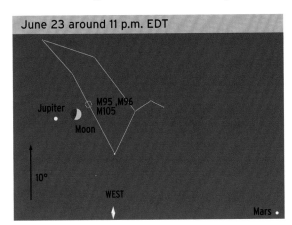

June 23 around 11 p.m. EDT

Jupiter

M95, M96
M105

Moon

10°

WEST

Mars

1 Tuesday		
2 Wednesday		
3 Thursday		■ 12:21 a.m. Full Moon ■ 9 a.m. Moon passes the perigee: 357,249 km
4 Friday		
5 Saturday		
6 Sunday		
7 Monday		
8 Tuesday		■ Venus crosses in front of the Sun 1:20 a.m. beginning 4:21 a.m. middle 7:24 a.m. end
9 Wednesday		■ 4:07 p.m. Last quarter of the Moon
10 Thursday		
11 Friday		
12 Saturday		
13 Sunday		
14 Monday		
15 Tuesday		

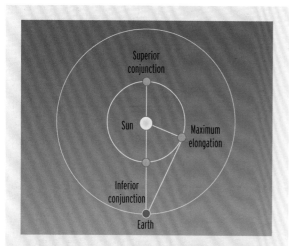

Elongation, conjunction, opposition

Mercury and Venus are permanently inside the Earth's orbit. They cannot move away from the Sun by an angle superior to that reached at their maximum elongation. Thus it is impossible to see them in the middle of the night, because they can never be observed on the opposite side to the Sun. When they are in the same direction as the Sun, they are said to be in conjunction. When the inferior planet is between the Earth and the Sun, there is an inferior conjunction, and when it is further away than the Sun, there is a superior conjunction. As for the superior planets visible to the naked eye, Mars, Jupiter and Saturn, observation is possible in the middle of the night. When the planet is opposite to the Sun, it is in opposition.

When it is in the direction of the Sun, it is in conjunction. In all cases, the superior planet in conjunction is further away than the Sun. On the other hand, people speak of a conjunction between two stars each time these two stars appear to move closer to one another.

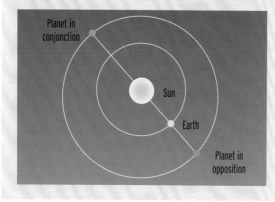

less than 1° from the Moon at this hour. You should resume observation a little later, above the western horizon: the Moon will then be between Jupiter and the group of galaxies, with all objects appearing in the field of a pair of binoculars.

Moving stars

During these early June evenings, look carefully at the apparent motion of the planets. It was thanks to their movement that they were identified as early as Classical times: the word planet comes from the Greek *planetes* which means "mobile" – the planets are therefore, etymologically, moving stars. They differ in this from all the stars that seem linked to one another in relatively unchanging positions, and which allowed us to draw the constellations. Only five "moving" stars are visible to the naked eye. If we add the Sun and the Moon, which are also moving, but look very different from the stars, we then have seven anomalies in the sky. The numbers five and seven, with their immediate connotations of magic and the sacred (the seven colors of the rainbow, the seven deadly sins, etc.), are thus engraved in the firmament.

The transit of Venus

Since it was established that the planets, including both the Earth and **Venus**, revolve around the Sun (they are heliocentric), Venus's transit in front of the Sun is a phenomenon that is possible to predict with the greatest accuracy. Curiously, the 1631 transit was not foreseen by Kepler, whereas the following one, in 1639, clearly figured among his predictions. These occurrences, while proving that the planet Venus really does go around the Sun, were also used to try to get a more

Observing the Sun by means of projection

This can be done without any risk with the help of a projection system developed at the French observatory of the Côte d'Azur, called the Solarscope®. It enables you to obtain an image of the solar disk about 4 inches (100 mm) across, under excellent safety conditions. This reasonably priced instrument ($60) is particularly suitable for observing an event like the transit of Venus.

accurate idea of the distance between the Earth and the Sun. Their rarity makes it an exceptional experience, recurring according to a complex cycle in which transits take place in pairs separated by eight years. The last ones happened in 1631 and 1639, 1761 and 1769, 1874 and 1882. After this year's, the next will be in 2012. So it is a rare phenomenon that has to be observed, but strictly in accordance with the precautions described on pp. 90–91: observing the Sun can be very dangerous for the eyes if very strict rules are not adhered to.

The event begins at 1:20 a.m. EDT with the first contact at the edge of the Sun's disk, and Venus is totally over the disk by 1:40 a.m. It takes a little less than 5 hours 25 minutes for the planet to pass in front of the Sun. Venus begins to emerge from the disk at 7:04 a.m. and the whole thing will have finished by 7:24. The time it takes will vary according to the observer's position. By comparing the duration of the transit measured at different points on Earth that, from 1761 onwards, astronomers tried to obtain more precise knowledge of the distance between the Sun and the Earth.

16 Wednesday
- The Moon and Venus move closer together. Visible at the end of the night (minimum 4.9° at 1 p.m.)

17 Thursday
- 2 p.m. Moon passes the apogee: 406,575 km
- 4:29 p.m. New Moon

18 Friday

19 Saturday
- The Moon and Saturn move closer together. Visible in the early part of the night (minimum 4.9° at 2 a.m.)

20 Sunday
- The Moon and Mars move closer together. Visible in the early part of the night (minimum 3.8° at 5 a.m.)
- 8:58 p.m. Summer solstice

21 Monday

22 Tuesday

23 Wednesday
- The Moon and Jupiter move closer together. Visible in the early part of the night (minimum 3.4° at 7 p.m.)

24 Thursday

25 Friday
- 3:11 p.m. First quarter of the Moon

26 Saturday

27 Sunday

28 Monday

29 Tuesday

30 Wednesday
- Maximum of the June Draconids meteor shower (2 meteors an hour)

A region of the sky to explore: Scorpius

The constellation of Scorpius is a zodiacal constellation, which is to say that the Sun, Moon and planets pass through it. If you add to it the stars making up the present constellation of Libra, the whole group really does resemble a scorpion. In Classical times, the stars of what is now Libra formed the constellation considered to be the claws of the adjoining scorpion. The Scorpion represents the animal that, according to one of the myths about Orion, killed the hero who had boasted of exterminating all the wild beasts on Earth. The story can be read in the sky: when Orion sets in the west, Scorpius rises in the southeast. Just above Scorpius is the constellation of Ophiuchus, the Serpent Bearer, representing Asclepius, the god of healing, in the process of defeating the scorpion.

The constellation of Scorpius is one of the most interesting to observe. It is an excellent indicator of the quality of the sky and you need to have a very clear southern horizon to be able to pick out the stars in the animal's tail. Several stars in the constellation came into being at the same time, about 20 million years ago, in a group of about a hundred members that form the Sco-Cen (Scorpius-Centaurus) association, described in 1914 by the Dutch astronomer Jacobus Kapteyn (1851–1922). You can find stars belonging to this association over most of the sky, at an average distance of around 500 light years. It is the nearest association of stars of this type, which are young, luminous and massive.

The rival of Mars

Antares (or **Alpha** (α) **Scorpii**), the most luminous star, is also the one with the greatest mass in the Sco-Cen association, and is particularly interesting. Its name, meaning anti-Mars, or the other Mars, was coined because its color is like that of the planet Mars and it is situated almost on the ecliptic ▶. It is a giant red star, about 423 light years away ▶ with a diameter 50 times that of the Sun. Antares, if it was in the same position as the Sun, would encompass all the planetary orbits as far as Jupiter.

DEFINITIONS ▶

Ecliptic: The plane that passes through the center of the Sun and contains the Earth's orbit.

Light year: The distance traveled by light in one year at a speed of 300,000 km/s. Expressed in kilometers, this distance is equal to 9,461 billion kilometers. The distance from the Earth to the Sun is about 500 light seconds.

The constellation of Scorpius

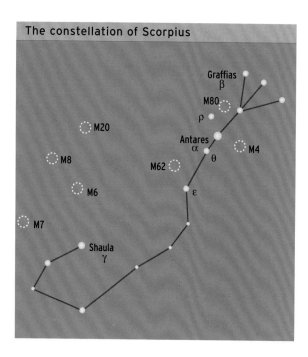

Graffias
β

M80

ρ

M20

Antares
α

M4

θ

M8

M62

M6

ε

M7

Shaula
γ

THE SKY IN JUNE

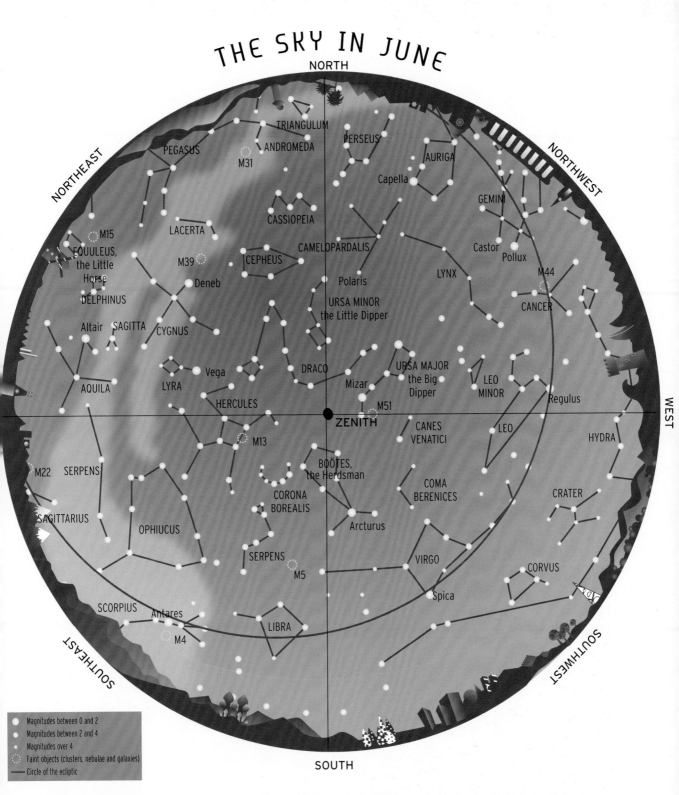

NORTH

NORTHEAST

NORTHWEST

TRIANGULUM
PEGASUS
ANDROMEDA
M31
PERSEUS
AURIGA
Capella
CASSIOPEIA
M15
LACERTA
GEMINI
EQUULEUS, the Little Horse
M39
CEPHEUS
CAMELOPARDALIS
Castor
Pollux
M44
DELPHINUS
Deneb
Polaris
LYNX
CANCER
Altair
SAGITTA
CYGNUS
URSA MINOR the Little Dipper
LYRA
Vega
DRACO
URSA MAJOR the Big Dipper
LEO MINOR
AQUILA
HERCULES
Mizar
M51
Regulus
WEST
M13
ZENITH
CANES VENATICI
LEO
HYDRA
M22
SERPENS
BOÖTES, the Herdsman
COMA BERENICES
CRATER
SAGITTARIUS
OPHIUCUS
CORONA BOREALIS
Arcturus
CORVUS
SERPENS
VIRGO
SCORPIUS
M5
Spica
Antares
LIBRA
M4

SOUTHEAST

SOUTHWEST

Magnitudes between 0 and 2
Magnitudes between 2 and 4
Magnitudes over 4
Faint objects (clusters, nebulae and galaxies)
— **Circle of the ecliptic**

SOUTH

■ **How to use this chart**
Hold the chart above your head, matching the word SOUTH that appears at the edge of it with the geographical south of the place you are observing from. Use a compass to help you do this.

■ **This chart shows the sky that is visible at a latitude of 45°**
If you are further north or further south, Polaris will be higher or lower in the sky.

■ **Chart of the sky visible at 11 p.m. EDT** at the beginning of the month; at 10 p.m. EDT in the middle of the month; at 9 p.m. EDT at the end of the month.

A month with your head up in the stars

▶ DEFINITION

Magnitude: The measure of a star's brightness (see pp. 110-11).

Antares is a pulsating star with a magnitude ▶ varying from 0.9 to 1.8 over a 53-month period. When it shines at its medium magnitude (of the order of 1), it is 6,300 times more luminous than our Sun. Together with Regulus, Aldebaran and Fomalhaut, Antares is part of the group known as "royal" in Classical times. Three of these first-magnitude stars are near to the ecliptic. Fomalhaut is not in the zodiac but in a region that is poor in luminous stars. These four royals served to divide up the sky and punctuate the four seasons, Antares marking the summer, Aldebaran the fall, Regulus the winter and Fomalhaut the spring.

Tau (τ) Scorpii is a beautiful bluish star with a magnitude of 2.8, situated about 300 light years away. Its surface temperature is on the order of 28,000 K, and its mass is about five times that of the Sun.

Rho (ρ) Ophiuchi is 3° to the north of Antares. It is a double star whose components are separated by 3.2". You can distinguish them with a small 2.5-inch (60 mm) diameter telescope; they are in a reflection nebula, a vast region of star formations. Immediately to the west of this pair of stars, it is possible, with the aid of a small instrument, to make out a black area, which is a dusty region concealing stars. In the 1780s, William Herschel described this obscure nebula as "a hole in the heavens." This vast hydrogen cloud has a mass of around 1,000 solar masses. It is a very interesting region to observe. On each side of Rho

Astronomical knowledge from Classical times has been handed down to us by the Arabs, who translated and continued the work of the astronomers of antiquity. This is why many stars still have Arabic names.

(ρ) Ophiuchi, forming a V shape, can be seen two stars with a magnitude of 7. The angular separation of this trio of stars is 2' 30".

Beta (β) Scorpii is also called Acrab or Graffias, perhaps from the Greek *graphaios*, "the crab." It is a beautiful double star, observable with a small instrument, and its components, with magnitudes of 2.6 and 4.9, are separated by 13.7". A companion with a magnitude of 9 is to be found 0.5" from the most luminous star. Acrab is about 600 light years away.

Dschubba (or **Delta** (δ) **Scorpii**) probably derives its name from the Arabic *al jabhah*, "the forehead." About 590 light years away, Dschubba is one of the most important members of the Sco-Cen association.

Lambda (λ) Scorpii, or Shaula, "the spear" in Arabic, has a magnitude of 1.62 and is one of the group of 25 brightest stars in the sky. This star is 35', that is about the diameter of the Moon, away from **Upsilon (υ) Scorpii**, Lesuth or Lesath, from the Arabic *al las'ah*, another form of spear. Lesath is another bright star (magnitude 2.71). For anyone living at latitudes further north than around 40°, it is not always very easy to observe these two stars with the naked eye, and it is better to use binoculars and try to determine their angular separation by this means.

The hunt for clusters

M4 (NGC 6121) is certainly one of the most beautiful globular clusters ▶ that can be admired with a small instrument. A close neighbor of Antares, it was observed by de Chéseaux in 1746 when

M4
This splendid globular cluster is close to Antares, in the constellation of Scorpius.

he was investigating comets, then in May 1764 by Charles Messier. Situated about 9,500 light years away, it is one of the globular clusters nearest to Earth. M4's main feature is that it is exceptionally rich and extensive, and this can be quite clearly seen through binoculars. However, in order to enjoy its beauty fully, it should be observed from low latitudes where it appears higher in the sky. It then reveals its exceptional qualities. You will find M4 by training your binoculars on Antares: it is within your field of vision, to the right, at about 1° 15', or a little more than twice the apparent diameter of the Moon.

▶ DEFINITION

Cluster: A group of stars linked by gravity. A cluster is called globular when the number and mass of stars are large enough for them all to become stabilized in a swarm. It is called open when the opposite is the case, with stars leaving the cluster in a few tens of millions of years (see p. 128)

M6 (NGC 6405) and **M7** (NGC 6475), situated 3° away from one another, are two open clusters to observe through binoculars. Ptolemy listed them in his catalog. M6 is 1,500 light years away and is easily found a little over 2° to the north of Gamma (γ) Scorpii, the star at the end of the Scorpion's tail. It is as large as the Moon and is

visible to the naked eye from the most southerly regions.

M7 is one of Charles Messier's discoveries (June 1771). Most of the stars in the cluster appear blue and hot, with the notable exception of the most luminous among them, which is orangy yellow.

M62 (NGC 6266) is a very dense globular cluster, whose asymmetry was noticed by John Herschel in 1847. It was cataloged by Messier in 1771. The cluster is quite difficult to observe with binoculars. On a very dark night, through a small telescope, its nucleus appears in the form of a beautiful disk. M62 is 37,000 light years away and contains a large number of variable stars.

M80 (NGC 6093) is a small, very luminous globular cluster that was discovered independently by Charles Messier and Pierre Méchain in January 1781. Through binoculars, you can make out a rather irregular disk. It is a very condensed cluster situated 36,000 light years away. To find M80, you just need to follow the straight line linking Antares and Graffias. The cluster is approximately halfway between the two stars.

M7
This open cluster is situated at the end of the constellation of Scorpius.

Practical astronomy

How to use a star chart

To find their way around the sky, observers need charts. A map enables you to get your bearings in the middle of the countryside but it has to be interpreted, since what is described is represented through symbols. Sky charts, on the other hand, depict what you can see. The stars always occupy the same relative positions, and the angles separating them do not change. The chart enables you to identify the main stars, recognize certain stellar configurations and foresee what the sky will be like in the future. Sometimes, the number of stars shown on the chart is greater than those visible with the naked eye. The chart thus appears more complicated than the visible night sky.

This is because it is intended for amateur astronomers who want to observe with the help of a telescope or binoculars. In very good weather, in areas far away from urban lighting and on a moonless night, it is possible to see about 3,000 stars. But from most of our cities, only a few dozen stars are visible.

A chart for each month

This book offers a chart of the sky for each month. Here is how to make best use of it.

Hold the chart above your head, so that the word NORTH appearing on the chart corresponds with the geographical north of your observation point.

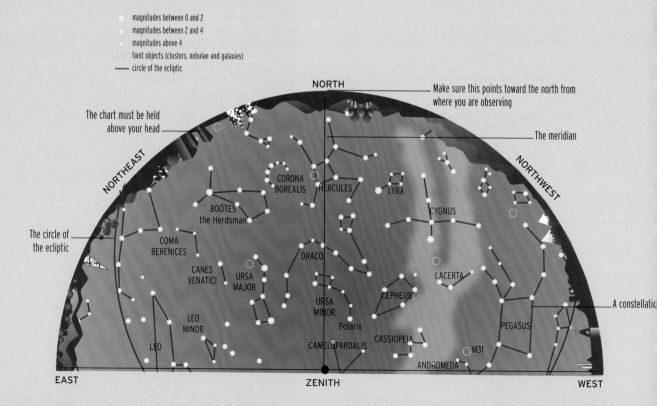

- ● magnitudes between 0 and 2
- ● magnitudes between 2 and 4
- · magnitudes above 4
- ○ faint objects (clusters, nebulae and galaxies)
- —— circle of the ecliptic

NORTH

Make sure this points toward the north from where you are observing

The chart must be held above your head

The meridian

NORTHEAST

NORTHWEST

CORONA BOREALIS HERCULES LYRA

CYGNUS

BOÖTES the Herdsman

The circle of the ecliptic

COMA BERENICES

DRACO

LACERTA

CANES VENATICI URSA MAJOR

CEPHEUS

A constellation

URSA MINOR

LEO MINOR

Polaris

CASSIOPEIA

PEGASUS

LEO

CAMELOPARDALIS

M31

ANDROMEDA

EAST

ZENITH

WEST

Help yourself with a compass in order to do this, and turn toward the north.

The chart represents the visible sky at a latitude of 45°. If you are further north or further south than this latitude, Polaris will be higher or lower in the sky. The chart shows the sky that is visible at a given hour on a given day – these are mentioned in the key. As the hours pass, you will see all the stars progressively turn from east to west. What is more, each month the sky shifts by two hours, or 30°. The sky seen at the beginning of June at 10 p.m. is like the sky at the beginning of July at 8 p.m. and the sky at the beginning of August at 6 p.m.

Finally, the chart shows the ecliptic, the circle representing the path of the Sun and the planets. The Moon never moves more than 5° from the circle of the ecliptic, passing through it twice a month, en route to the nodes of its orbit. If you prefer to use your computer, there are many types of software available on the Internet giving detailed charts of the sky for any date and place. Addresses and references are given at the back of this book (see p. 227).

Adjustable star charts (planispheres)

Once you are familiar with sky charts, you will be capable of using an adjustable chart or planisphere, enabling you to view the sky at any moment. To use this you need to know the date, time, and orientation of your observation point. Begin by bringing together the mobile markers showing the day and time when you intend to observe. These markers appear on the circles around the edge of the chart. Make sure you use Greenwich Mean Time, which is five hours ahead of Eastern Standard Time, eight hours ahead of Pacific Standard Time. Then hold the chart above your head, facing north, as you did with the fixed chart. Some adjustable charts allow you to locate the planets among the stars. They have a graduated ecliptic circle, usually in hours. You just need to use the attached ephemeris, which gives you the ecliptic longitude of the planet according to the date. If you mark the points indicated by the ephemerides on the ecliptic circle of your chart, you will locate the planets in the sky and easily find them again, as well as being able to predict them.

Augular separation
Here is a simple method of assessing the angles separating two stars: making sure you always have your arm fully extended, an open hand with fingers spread out covers 15°, a closed fist 10° and the index finger approximately 1°.

A cloud of stars
Filaments of interstellar matter are lit up by an open star cluster (Hodge 301). This Tarantula Nebula formation belongs to the Large Magellanic Cloud.

Discovering the stars

For the observer, stars are characterized by their position in the sky, their brightness and their color.

Position

The position of each star is always changing. The Earth's rotation makes the stars move in an apparently circular way, causing them to revolve around the sky in a single day. Another noticeable change, caused by the Earth turning around the Sun, is that of the stars shifting from day to day towards the west: An observer looking at the sky each night at the same time will notice that all the stars shift westwards by about 1° per day. The two movements do not affect the stars' relative positions, and the figures formed by the most visible ones, the constellations, remain unchanged. Thus, for the astronomer, each star has a known position, and coordinates which allow it to be observed at a given moment from a given place. Each one (of those that are most visible, like Arcturus, Antares, 61 Cygni, etc.) has a name or an identifier (Wolf 359, L276–8, etc.).

Brightness

It is easy to see that not all stars have the same brightness. From Classical times onwards, the stars were classified by "magnitude." The most luminous were said to be of first magnitude, the others of second, third, and so on, up to the sixth magnitude, which grouped together stars that were hardly perceptible. This empirical classification was replaced by a more objective system based on scientific measurement of a star's brightness, called the "scale of magnitudes." So as not to upset people's ideas and habits too much, this system of magnitudes was modeled exactly on the Classical magnitudes. The more luminous a star is, the lower its magnitude. A star with a magnitude of 1 is 2.5 times as luminous as a star with a magnitude of 2. The scale was then extended to negative values, and after the advent of optical instruments it became necessary to classify stars that were invisible to the naked eye at magnitudes well above 6. The 20 stars that Hipparchus put in the class of first magnitude do not in fact have the same brightness. Their "measured" magnitudes range from –1.4 for Sirius, the most luminous of all, to 0 for Rigil Kentaurus, in the constellation of Centaurus. With giant instruments, astronomers can record light from stars with a magnitude of more than 25, that is 40 billion times less luminous than Sirius. The apparent brightness of such a star is equivalent to the light from a candle placed as far away as the Moon!

Apparent magnitude corresponds to the amount of energy received on Earth, which is different from the amount of energy actually emitted by the star – two identical stars situated at different distances do not have the same apparent magnitude. Astronomers compare stars by calculating the brightness they would have if they were all situated at the same distance from the Earth. The distance chosen is 10 parsecs ▶, that is to say just over 32 light years. The absolute magnitude of a star is so called because of the magnitude it would have if it were situated at this standard distance.

Color

Another parameter to describe stars, which is directly accessible to the observer, is color. As is easy to see when observing the sky, the stars are of different colors. Some are a deep red (Antares), others are orange red (Arcturus) and some bright blue (Vega). The color of stars is a precise indicator of their outer temperature. The hotter an object is, the bluer the light it emits. Unlike the codes commonly used in our daily lives, where red symbolizes heat and blue cold, physics teaches us that any red body is colder than a blue body. The blue star Vega is therefore hotter than the red star Antares: 10,000 K for the one against 3,000 K for the other, which is a big difference. The surfaces of the hottest stars reach temperatures of several tens of thousands of degrees. "Cold" stars only have a surface temperature of 3,000 K to 4,000 K, which is still much more than the temperature of fire that we are familiar with. We should not, therefore, compare stars to fireballs. They are much hotter. Stars are very hot balls of gas and the temperature at the core can reach several tens of millions of degrees.

Introduction to astronomy

Distance

In order to classify and to get to know and understand stars, it was necessary to measure their distances away from us. This is one of the most difficult tasks there is, and astronomers had to wait until the middle of the 19th century to be able to measure the first ones. The classification of stars has been one of the major preoccupations of astronomers for more than 100 years. We now know that all the stars visible to the naked eye are near to Earth. They belong to the solar area of our galaxy, the Milky Way. The furthest visible stars are several thousands of light years away. The nearest star, on the other hand, is very distinctive. It is a triple star, of which the component nearest to the Sun, Proxima, turns around a pair of stars in the space of a few million years, and these stars in turn revolve around one other over an 80-year period. At present, Proxima is 4.34 light years away from the Sun.

The life of stars

Thanks to numerous classifications – by spectral type (linked to color), by intrinsic brightness if it has been possible to measure distances – it is possible to draw up statistics for stars and to reconstruct their history. The way this is done is like piecing together one person's life from photographs of a host of individuals chosen to represent all ages and all possible types of existence. Astronomers thus have the possibility of reconstructing the scenario of a star's evolution.

All stars are born in groups, from an immense cloud of interstellar matter; a few hundred are formed and begin evolving. Some remain linked to one another by gravity, in pairs or in more complex systems. This is why there are many double or triple stars, as well as associations with several dozen members. The length of a star's life appears to be closely linked to its size – the ones with a greater mass rapidly consume their hydrogen and disappear within a few tens of millions of years, at the moment when stars with a smaller mass are only just starting to reach adulthood and will continue living for billions of years. For small stars, like the

The most luminous stars

Name	Constellation	Apparent brightness	Absolute brightness
Sirius	Canis Major	−1.44	0.7
Canopus	Carina	−0.62	−5.5
Arcturus	Böotes	−0.05	−0.3
Rigil Kentaurus	Centaurus	−0.01	4.6
Vega	Lyra	0.03	0.3
Capella	Auriga	0.08	0.1
Rigel	Orion	0.18	−7
Procyon	Canis Minor	0.40	−1.3
Betelgeuse	Orion	0.45	−5.5
Achernar	Eridanus	0.45	−1.3
Agena	Centaurus	0.61	−4.3
Altair	Aquila	0.76	2.3
Alpha Crucis	The Southern Cross	0.77	−3.8
Aldebaran	Taurus	0.87	−0.2
Spica	Virgo	0.98	−3.2
Antares	Scorpius	1.06	−4.5
Pollux	Gemini	1.16	0.7
Fomalhaut	Pisces Austrinus	1.17	1.8
Deneb	Orion	1.25	−7.2
Mimosa	The Southern Cross	1.25	−4
Regulus	Leo	1.36	−1

A globular cluster
The Omega Centauri cluster (central area seen here) is one of the most beautiful globular clusters. It is visible with the naked eye from the southern hemisphere.

The nearest stars			
Name	Constellation	Distance in light years	Apparent brightness
Proxima, α Cen C	Centaurus	4.34	11.01
α Cen B	Centaurus	4.39	1.35
Rigil Kentaurus, α Cen A	Centaurus	4.39	−0.01
Barnard's Star	Ophiuchus	5.93	9.54
Gliese 411	Ursa Major	8.31	7.49
Sirius A and B	Canis Major	8.60	−1.44
Gliese 729	Sagittarius	9.70	10.37
ε Eridani	Eridanus	10.51	3.72
Gliese 887	Pisces Austrinus	10.75	7.35
Ross 128	Virgo	10.90	11.12
61 Cygni A	Cygnus	11.35	5.20

Sun, evolution lasts for about ten billion years, which are spent transforming the central hydrogen into helium to compensate – through energy radiation – the forces of gravity which have a tendency to make matter collapse. When evolution has finished, and the quantity of central hydrogen is no longer sufficient, the star ejects the matter it has preserved and transformed during the whole of its evolution, into the interstellar environment. Cores of helium are synthesized from hydrogen, then oxygen, carbon and silicon, as the temperature at the center increases during its final evolutionary stages. All atoms existing in the Universe, other than hydrogen, helium and lithium, have been formed during the course of these transformations undergone by the stars.

We are the living proof that stars evolved before the Sun existed. Without them, the heavy atom content – carbon, oxygen, sodium, calcium, etc. – in the primitive nebula from which the solar system emerged would not have allowed cells as complex as ours to be produced. Information about stars is now accumulating and several hundred thousand stars are being cataloged, followed and measured all the time. New classifications are being developed, and new discoveries are extending this quest for knowledge that began 2,600 years ago and led to the discovery that our world was no more than a planet revolving around the Sun, that the Sun itself was an ordinary star among a hundred thousand others grouped in our galaxy, and that this in turn was only one of billions of other galaxies.

Stars are only suns, and the Sun is no more than a small star. Since 1995, we have known that the Sun is not the only star to possess planets. The big question now is whether, on these other distant planets, or "exoplanets," life is possible – and if it already exists. Unfortunately, the number of stars is so vast, that we may never have the answer.

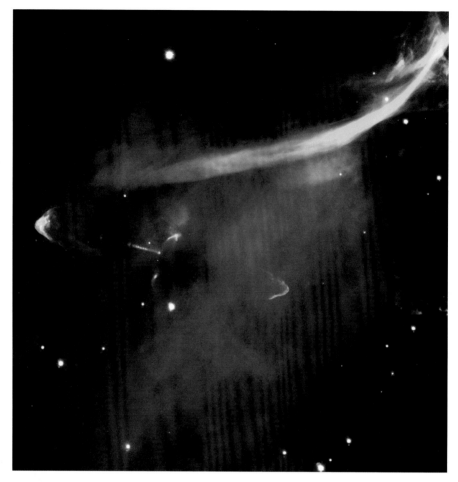

In the vicinity of a protostar
It is from a planetary system in the process of formation, like HH34 in the constellation of Orion, seen here, that protoplanets will be born.

The hunt for exoplanets

In 1995, for the first time astronomers detected a planet revolving around a star other than our own. Our solar system is therefore not unique. We are living through a revolution as important as the one at the beginning of the 17th century, when it was discovered that our world was not the only one and the Earth was just one planet amongst others. Giordano Bruno was one of the first to assert that the stars were suns and that they might be surrounded by numerous worlds. The observation equipment used up until the end of the 20th century did not allow us to discern these obscure companions drowned in starlight. With the stars being such enormous distances away, the planetary systems are seen from infinitesimal angles. Thus, for an observer on Proxima, our nearest star, the angle separating the Sun from Jupiter is an arc of only a few seconds, whilst their difference in brightness is of the order of a billion. With present-day equipment, it is therefore impossible to observe an exoplanet directly. Perhaps, with the rapidity of technological development, this will change in a few years.

The first

In fact, the first exoplanets were detected by observing the movements they cause in the star that they are orbiting. It was by precise measurement of the star's speed of movement along the line of sight, or its radial velocity, that it was possible to identify these exoplanets. Research is concentrated on solar-type stars, whose spectra have a very large number of lines that can be precisely measured. Michel Mayor and Didier Queloz, Swiss astronomers using the 193-cm telescope at the observatory in Haute-Provence in France, were lucky: 51 Pegasi, the star whose radial velocity they were recording with an instrument called Coralie, was showing periodic variations of only a few days. But this was enough to reveal the existence of the first exoplanet, which had some astonishing features: it was in fact a giant planet, of the same kind as Jupiter, revolving around its star in a little over four days! This rapid rotation facilitated the discovery because, in order to detect a planet as far away as Jupiter, about 10 years of observation would be necessary! Since 1995, the technique for detecting exoplanets by measuring radial velocity has enabled several dozen planetary systems to be identified, together with the position and approximate mass of the perturbing planet in each case. The improvement in techniques for sifting data and the increase in the number of measurements is leading to a rapid increase in these discoveries. It will still be a very long time before more concrete information about the nature of these planetary systems can obtained, and whether they are likely to support life.

One of the four telescopes that make up the VLT

When we will see exoplanets?

When will we see exoplanets?

Other detection techniques can be put to work. One of them consists of precisely measuring the variations in a star's brightness. If a planet passes between the star and the observer, the reduction in brightness caused by the partial eclipse can be recorded. This technique, combined with the variation in radial velocity, enables the mass and diameter of the planet that has been detected to be ascertained precisely. Other techniques will emerge with the appearance of new observation equipment. The dream of astronomers is to be able to observe these exoplanets directly. The arrival of giant telescopes, like the VLT (Very Large Telescope - see p. 151), with a better power of resolution, will soon enable them to view these faint planets, which are very close to their star. Several planets with periods of a few years have already been located. The record is at present held by one with a period of more than eight years - perhaps a real Jupiter! The ambition at first is to observe planets as small as the Earth. Then scientists will photograph and analyze their surface and atmosphere, possibly picking up signs of biochemical activity - in other words, signs of life.

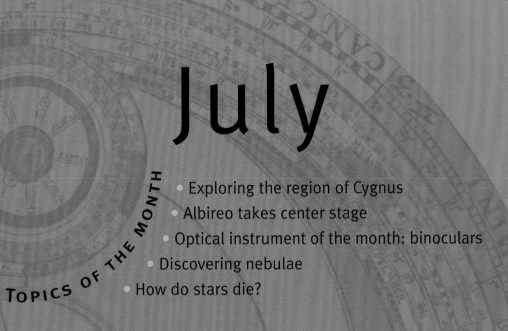

July

NGC6369, the cosmic ghost
This halo is formed by the remains of an exploding star, whose skeleton can be found in the center of the image.

A month with your head up in the stars

The sky is dominated by the Summer Triangle, made up of Deneb, Vega and Altair - three lovely stars that are very bright and easy to find. Jupiter, still visible in the west at the beginning of the month, will disappear in the Sun's light while moving nearer to Regulus, the brightest star in the constellation of Leo.

Distant galaxies
One of these galaxies, the larger one, NGC4319, is 80 million light years away. The other, Markarian 205, is a quasar about one billion light years away.

Observations of the month

In the night of July 4–5, the Earth is at the aphelion, which means it is at its maximum distance from the Sun. The phenomenon of the seasons is therefore not caused by a variation in the distance between Earth and Sun, but by the inclination of our planet's rotation axis in relation to the plane of its orbit. Ursa Major continues to move downward toward the northern horizon: at about 10 p.m., the handle seems to be clinging to the horizon. The Summer Triangle dominates and the first star to appear, very high in the sky, is the blue-colored Vega, in the constellation of Lyra. This is the time to look again, through binoculars, at the constellations of Libra, Scorpius and Sagittarius on the southern horizon. Some beautiful non-stellar objects can easily be seen.

All times are given in Eastern Daylight Time (EDT). Calculations have been made for a northern latitude of 44° and a western longitude of 80°.

TIMES OF RISING AND SETTING FOR THE SUN, MOON AND FIVE PLANETS VISIBLE WITH THE NAKED EYE													
Date	Sun		Moon		Mercury		Venus		Mars		Jupiter		Saturn	
	Rises	Sets	Rises	Sets	Rises	Sets	Rises	Sets	Rises	Sets	Rises	Sets	Rises	Sets
7/1	5:43 a.m.	9:04 p.m.	9:07 p.m.	5:19 a.m.	6:44 a.m.	10:06 p.m.	4:00 a.m.	6:28 p.m.	7:43 a.m.	10:38 p.m.	11:08 a.m.	12:14 a.m.	6:12 a.m.	9:27 p.m.
7/5	5:45 a.m.	9:03 p.m.	11:38 p.m.	9:18 a.m.	7:08 a.m.	10:14 p.m.	3:48 a.m.	6:14 p.m.	7:41 a.m.	10:29 p.m.	10:55 a.m.	12:00 a.m.	5:59 a.m.	9:30 p.m.
7/10	5:48 a.m.	9:01 p.m.	1:16 a.m.	3:09 p.m.	7:34 a.m.	10:17 p.m.	3:34 a.m.	6:01 p.m.	7:38 a.m.	10:18 p.m.	10:39 a.m.	11:42 p.m.	5:43 a.m.	8:56 p.m.
7/15	5:53 a.m.	8:58 p.m.	3:46 a.m.	8:37 p.m.	7:57 a.m.	10:16 p.m.	3:22 a.m.	5:51 p.m.	7:35 a.m.	10:07 p.m.	10:24 a.m.	11:24 p.m.	5:26 a.m.	8:38 p.m.
7/20	5:58 a.m.	8:54 p.m.	8:52 a.m.	10:37 p.m.	8:15 a.m.	10:09 p.m.	3:12 a.m.	5:45 p.m.	7:32 a.m.	9:56 p.m.	10:09 a.m.	11:06 p.m.	5:10 a.m.	8:21 p.m.
7/25	6:03 a.m.	8:50 p.m.	2:44 p.m.	12:19 a.m.	8:27 a.m.	9:58 p.m.	3:04 a.m.	5:41 p.m.	7:29 a.m.	9:45 p.m.	9:54 a.m.	10:49 p.m.	4:53 a.m.	8:04 p.m.
7/30	6:08 a.m.	8:45 p.m.	8:39 p.m.	4:06 a.m.	8:33 a.m.	9:44 p.m.	2:58 a.m.	5:39 p.m.	7:26 a.m.	9:33 p.m.	9:39 a.m.	10:31 p.m.	4:37 a.m.	7:43 p.m.

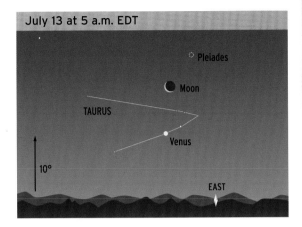

July 13 at 5 a.m. EDT

Pleiades

Moon

TAURUS

Venus

10°

EAST

The planets

A spectacular sight, early in the morning of the 13th, will be the **Moon** and **Venus** moving toward one another in the constellation of Taurus, with the Pleiades to the west of the crescent Moon. The 21st, at 9 p.m., will also offer a good opportunity for observation, with **Jupiter**, Beta (β) Leonis and the crescent Moon aligned in that order on 5°, and **Mercury** and **Mars** low on the western horizon.

▶ DEFINITION

An astronomical unit (AU): the mean distance of the Earth from the Sun, or 149,597,870 km.

You should try to observe Jupiter for as long as possible. Each evening, it moves closer to the Sun, but is visible before nightfall if weather conditions are

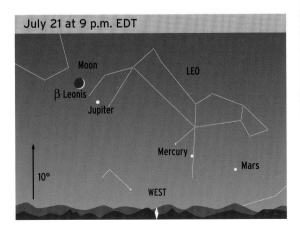

July 21 at 9 p.m. EDT

Moon

LEO

β Leonis

Jupiter

Mercury

Mars

10°

WEST

1 Thursday		■ 7 p.m. Moon passes the perigee: 357,450 km
2 Friday		■ 7:10 a.m. Full Moon
3 Saturday		
4 Sunday		
5 Monday		■ 1 a.m. The Earth passes furthest from the Sun (aphelion): 152,096,154 km
6 Tuesday		
7 Wednesday		
8 Thursday		
9 Friday		■ 3:03 a.m. Last quarter of the Moon
10 Saturday		
11 Sunday		
12 Monday		
13 Tuesday		■ The Moon and Venus move closer together. Visible at the end of the night (minimum 7.7° at 8:10 p.m.)
14 Wednesday		■ 5 p.m. Moon passes the apogee: 406,192 km
15 Thursday		

good and the western horizon is cloud-free. Venus can be clearly seen at the end of the night and, right at the end of the month, **Saturn** is beginning to appear early in the morning.

Shooting stars

Shooting stars will be quite plentiful this month, but conditions will not be very good for observing the **Omicron** (o) **Draconids**, which will be masked by the light of the Moon. They are the remnants of the comet Metcalf, discovered in 1919. The most active shower of the month is the **Southern Delta** (δ) **Aquarids**, which lasts from July 14 to August 28, with a maximum on July 28 and 29. These shooting stars appear to come from the comet Honda-Mrkos-Pajdusakova, discovered in 1948, which has a period of 5.25 years, and was last observed in the year 2000.

A region of the sky to explore: Cygnus

During these summer months, it is possible to make out a large cross hanging over our heads. In the Middle Ages, this was known in Europe as the Northern Cross, and was a mascot for actors, who saw in it a sign of good luck. It has to be said that plays in those days took place in the open air, so it is understandable that people came in greater numbers in summer when the weather was good and the Northern Cross was therefore more visible. This enormous cross, bounded by the stars Deneb and Albireo on its longest side, is also reminiscent of a huge bird, and for this reason it was christened "the Swan" in Classical times. Among the numerous legends associated with this constellation, the most important concerns Leda. Jupiter transformed himself into a swan in order to seduce her, and she then gave birth to twins who were to experience very different fates – Pollux was immortal, unlike his brother Castor. This is why Gemini, the twins, are not visible throughout the year – when the Swan–Jupiter dominates the sky, Pollux goes away in sadness at Castor's descent into hell, and the two brothers are invisible. It is only when the Swan disappears over the northwest horizon that the twins appear in the northeast.

Deneb (or **Alpha** (α) **Cygni**), derives its name from *al dhanab al dajajah*, "the hen's tail." Several stars have the name Deneb, which generally means the tail of an animal, e.g., Denebola ("lion's tail"), Deneb al Kaita ("whale's tail"). Deneb Cygni forms the summit of the Northern Cross, and is the most northerly of the stars in the Summer Triangle. It is one of the biggest known super-giants, its luminosity being equal to 60,000 times that of the Sun. Its mass, which is 25 times greater than the Sun's, puts it among the large stars likely to generate a pulsar at the end of its evolution. Its distance is in the order of 1,600 light years and its outer temperature is 9,700 K. Deneb is a giant star, with a diameter 60 times that of our Sun. It is also slightly variable, with a magnitude of

M29
This open cluster in the constellation of Cygnus can be seen with small instruments.

0.05. This is probably due to the star's pulsations or to large-scale turbulence in its atmosphere.

Albireo the show-stealer

Albireo (or **Beta** (β) **Cygni**) is the swan's head. Its original name was *al minhar al Dajajah*, "the hen's beak." It is a sumptuous double star, probably a physical double, which is to say the pair is linked by mutual gravitational attraction. However, their period is still unknown.

The distance from Albireo to the Earth is about 410 light years. The two components are of very different color – topaz yellow and sapphire blue – corresponding to a wide variation in temperature. The blue star is very hot (about 20,000 K) whereas the yellow star does not even reach 7,000 K. Albireo provides an excellent observation test: hardly visible to the naked eye, it reveals its splendors when viewed through different optical instruments. With binoculars, you can clearly make out the two components, whereas with a 2.5-inch (60 mm)

16 Friday
- The Moon and Saturn move closer together. Visible at the end of the night (minimum 5° at 3:26 p.m.)
- Maximum of the Omicron (o) Draconids meteor shower (3 meteors an hour)

17 Saturday
- 7:25 a.m. New Moon

18 Sunday
- The Moon and Mars move closer together. Visible in the early part of the night (minimum 5° at 9:57 p.m.)

19 Monday
- The Moon and Mercury move closer together. Visible in the early part of the night (minimum 5° at 11:19 a.m.)

20 Tuesday

21 Wednesday
- The Moon and Jupiter move closer together. Visible in the early part of the night (minimum 1.9° at 9:21 a.m.)

22 Thursday

23 Friday

24 Saturday
- 11:38 p.m. First quarter of the Moon

25 Sunday

26 Monday
- 11 p.m. Greatest eastern elongation of Mercury (27.1°)

27 Tuesday

28 Wednesday
- Maximum of the Southern Delta (δ) Aquarids meteor shower (20 meteors an hour)

29 Thursday

30 Friday
- 2 a.m. Moon passes the perigee: 360,326 km

31 Saturday
- 2:06 p.m. Full Moon

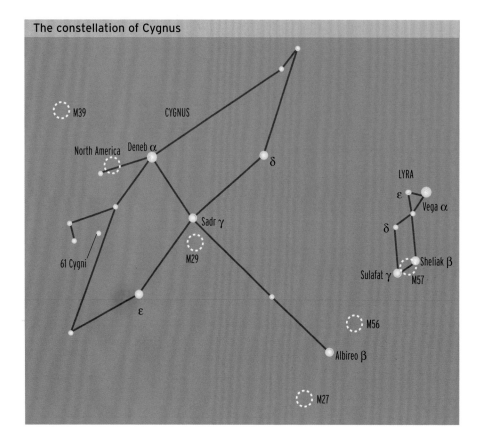

The constellation of Cygnus

diameter telescope, the colors become more distinctive and the contrast is striking.

Color and temperature are connected. The hotter a body is, the more it radiates in short wavelength domains: when it is heating up, it starts by radiating in the infrared, then it reddens and becomes more and more luminous in the visible domain, going from red to blue with the increase in temperature. If the temperature continues to rise, the radiation moves toward the ultraviolet, then toward X-rays and gamma rays. Red stars, like Antares, are the ones that are least hot, whereas blue stars like Vega are much hotter. Train yourself to assess the surface temperature of stars by observing their color. The night sky in summer lends itself particularly well to this exercise. Try observing Vega (10,000 K), Antares (3,000 K), Arcturus (4,500 K) and others. It is not so difficult to get an idea of stars' temperatures when you know how to look at them.

Sadr (or **Gamma (γ) Cygni**), from the Arabic *al sadr al dajajah*, "the chicken's breast," is situated 750 light years away. Its luminosity is equal to 5,800 times that of the Sun.

61 Cygni is one of the most important stars in the history of astronomy. Visible with small instruments, it is a beautiful double star whose components are slowly rotating. Its period is seven centuries. It became famous in 1792, when Giuseppe Piazzi detected its very substantial movement. He called it "the flying star."

THE SKY IN JULY

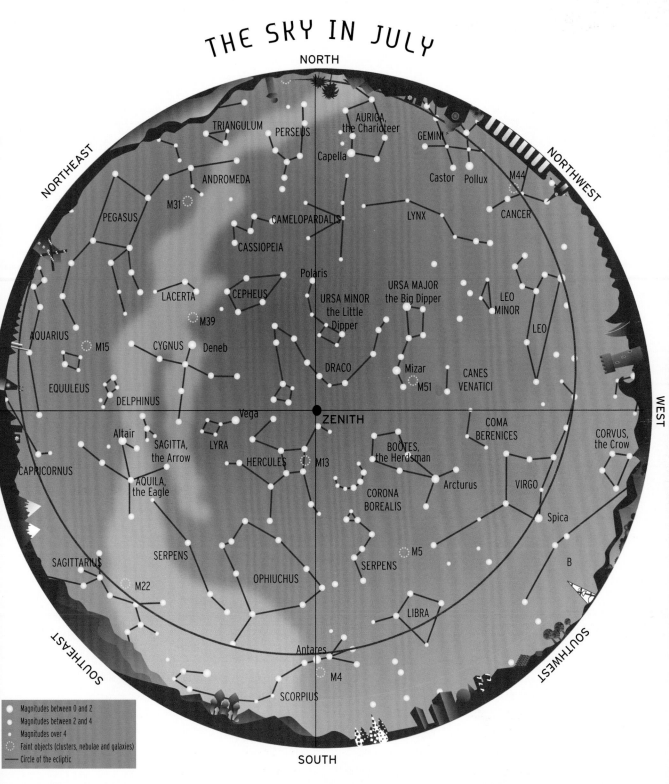

NORTH

TRIANGULUM
PERSEUS
AURIGA, the Charioteer
GEMINI
Capella
Castor Pollux
M44
ANDROMEDA
M31
CANCER
PEGASUS
CAMELOPARDALIS
LYNX
CASSIOPEIA
LEO MINOR
LACERTA
CEPHEUS
Polaris
URSA MAJOR the Big Dipper
LEO
M39
URSA MINOR the Little Dipper
AQUARIUS
M15
CYGNUS Deneb
DRACO
Mizar
CANES VENATICI
M51
EQUULEUS
DELPHINUS
COMA BERENICES
Altair
Vega
ZENITH
CORVUS, the Crow
SAGITTA, the Arrow
LYRA
BOÖTES, the Herdsman
CAPRICORNUS
HERCULES
M13
Arcturus
VIRGO
AQUILA, the Eagle
CORONA BOREALIS
Spica
B
SAGITTARIUS
SERPENS
M5
SERPENS
M22
OPHIUCHUS
LIBRA
Antares
M4
SCORPIUS

NORTHEAST **NORTHWEST** **WEST** **SOUTHWEST** **SOUTHEAST**

SOUTH

- ○ Magnitudes between 0 and 2
- ○ Magnitudes between 2 and 4
- · Magnitudes over 4
- ○ Faint objects (clusters, nebulae and galaxies)
- — Circle of the ecliptic

■ **How to use this chart**
Hold the chart above your head, matching the word SOUTH that appears at the edge of it with the geographical south of the place you are observing from. Use a compass to help you do this.

■ **This chart shows the sky that is visible at a latitude of 45°**
If you are further north or further south, Polaris will be higher or lower in the sky.

■ **Chart of the sky visible at 11 p.m. EDT** at the beginning of the month; at 10 p.m. EDT in the middle of the month; at 9 p.m. EDT at the end of the month.

Parallax: If you look at near objects closing each eye one after the other, they seem to move in relation to the background; in the same way, stars which are near to us appear to move at six-monthly intervals in relation to more distant stars, because the position of the Earth is changing due to its movement around the Sun. The angle of this movement is the parallax.

For the closest stars, it only measures a fraction of a second of arc (the angle at which you can see a person 300 km away).

In 1838, Friedrich Wilhelm Bessel chose it for his first attempt at detecting parallax ▶. Bessel established a parallax of 0.29", that is a distance of 10.3 light years. The distance now accepted is 11.01 light years, putting it at fourth place in the distance hierarchy of stars visible to the naked eye, after Alpha (α) Centauri, Sirius, and Epsilon (ε) Eridani.

The North America Nebula
NGC 7000 owes its name to its shape, recalling that of the North American continent.

Nebulae and clusters

M29 is a small open cluster discovered by Charles Messier in July 1764. The object is visible through binoculars, but quite tricky to find in this densely populated region. It can be spotted thanks to the attractive quadrilateral formed by four of its components. The density of dust present in the cluster is about 1,000 times greater than the average amount found in our galaxy. M29 is about 3,000 light years away and contains numerous red stars.

M39, a large open cluster discovered by Guillaume Legentil in 1750, is situated about 800 light years away. It is possible that it was mentioned by Aristotle in about 325 BC, as being a comet-like object. It is particularly beautiful when seen through 10 x 50 binoculars, which enable you to distinguish ("resolve" in astronomical language) its main components.

NGC 7000, the large nebula in the shape of North America, is often, in fact, called the North America Nebula by amateur astronomers. It was Max Wolf who, having discovered it via photography, gave it this name. The cloud is illuminated by Deneb and is about 1,600 light years away. It is difficult to make it out through binoculars or a telescope because of the lack of contrast. This is why it does not appear in Charles Messier's catalog. But this object is the delight of photographers – even with rudimentary instruments it is possible to obtain magnificent pictures of this nebula. America is situated 5° to the southeast of Deneb.

© Philip Perkins 1998

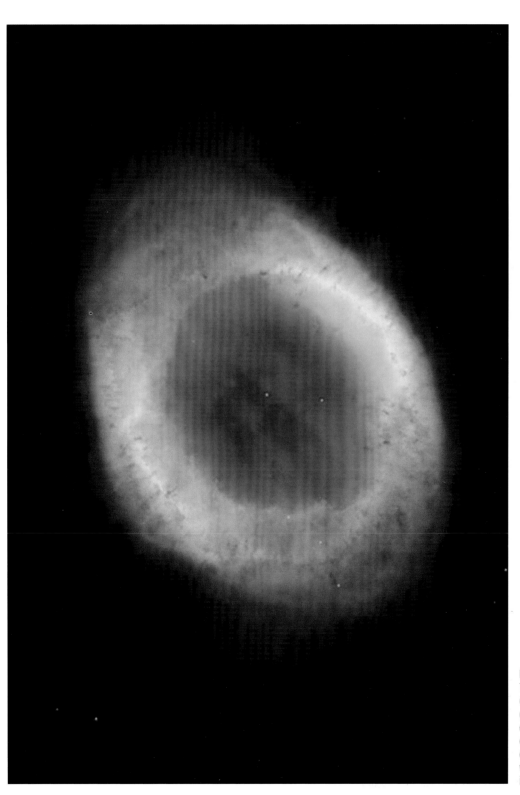

M57
The ring nebula of Lyra, the great attraction of summer evenings, can easily be seen with a small instrument.

Practical astronomy

Optical instrument of the month: binoculars

Binoculars are remarkable astronomical instruments for discovering the hidden wonders of the sky. The diameter of each of the lenses allows them to receive about a hundred times more light than our pupils, so they reveal objects that we are incapable of seeing with our eyes, making these a hundred times brighter and bringing them ten times nearer. Binoculars reveal details that we could only see by reducing the distance separating us from them by 90%. The sky observed through binoculars is a remarkable and fascinating sight.

How do I adjust them?

Generally, our two eyes are not identical. You therefore need to adjust each of the lenses to accommodate this slight difference. Usually the right lens is adjustable and the left one is fixed. To adjust, you should begin with your left eye: cover the right-hand lens with the lens cap or your right hand, then sharpen up the clarity by turning the focusing ring in the center of the binoculars. When the image is clear, cover the left-hand lens and adjust the view through the right eye by turning the right-hand lens. Then you just need to decide how much space you want between the eyes in order to obtain a single clear binocular image.

How do I choose them?

All binoculars carry two figures separated by an "x" sign, e.g., 7 x 35, 7 x 50, 8 x 40 or 20 x 80. The first indicates the degree of linear magnification or "bringing closer" (seven times closer in the first example). The second is the diameter, in millimeters, of each of the lenses.

As with all light collectors, the essential feature is the diameter. The amount of light gathered is in fact proportional to the diameter of the opening. For nighttime observation, you must choose a diameter equal to at least 30 mm; 50 mm is a good diameter for the purposes of astronomy because, beyond that, the binoculars are less comfortable on account of their weight and awkwardness. It is not advisable to use over-strong magnification for everyday observation. If it magnifies over ten times, the use of a stand is essential. Adapters will allow you to place the binoculars on a photographic tripod or a stand with a single leg.

There are two main types of binoculars: prism binoculars and direct binoculars, the former being more compact. In both cases, an optical system straightens the image. You can simply test the lens transmission quality by holding the binoculars with your arm outstretched

The moon
The great lunar formations become apparent through binoculars.

and looking at the shape of the bright patch which can be seen on each lens. If these patches are perfectly round, the lens has been treated so as to guarantee excellent transmission. If the patches are angular, generally in the form of a square or a hexagon, transmission will not be perfect at the edges of the field, due to a loss of peripheral brightness. This detail poses a handicap to viewing, and explains price differences, which essentially correspond to the difference in the way the lenses are treated.

Prices

The prices of binoculars can vary a great deal. You will find good quality instruments for about $100 and excellent products for between $100 and $700. There are also monsters with outstanding features (40 x 150), but prices then become literally astronomical, going well beyond the $10,000 mark. Stabilizing binoculars, which incorporate a system to absorb any shaking by the person holding them and restore a perfectly still image, are very comfortable to use. You can buy a good pair for around $400.

What should you observe?

The first thing to observe through binoculars is the Moon. Choose a full Moon for maximum relief and effect. Remember, though, that the Moon is not near enough for all its details to be visible. This is when strong magnifying lenses are particularly beneficial: 15 x 40 or 20 x 60 give excellent results, but need a stand to support them. If you have more traditional binoculars, 7 x 50 for example, be careful not to miss the Moon's eclipses, when the Earth's shadow moving over the larger craters provides a fascinating spectacle. It is also interesting to determine the shadow's color, which is an indicator of pollution in the Earth's atmosphere.

Binoculars also provide an excellent way of observing occultations of bright stars and planets. Jupiter and Saturn can easily be identified with binoculars. Jupiter's four Galilean satellites are visible even with small binoculars.

It is not easy to see Saturn's ring without good stabilization. But binoculars cannot be bettered when it comes to observing very large objects like clusters - e.g., Hyades, Pleiades and the Perseus double cluster - or close galaxies like Andromeda, M81.

Another very satisfying experiment with binoculars is to settle yourself comfortably and take off on a celestial tour with no other aim but to discover, and try to identify, the starry population of our sky that is invisible to the naked eye. Certain regions are particularly rich and unusual. In summer, try going toward the south, to the region of Sagittarius. If it is a fine night, you will find dozens of remarkable and unsuspected objects there.

The Pleiades
(far right)
Binoculars enable you to distinguish the stars in this superb cluster.

The Perseus double cluster
(right)
Through binoculars, nebulosities are seen to be star clouds.

Introduction to astronomy

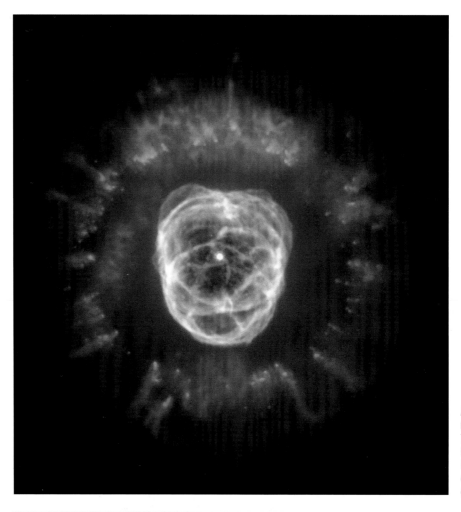

A planetary nebula
The Eskimo Nebula is found in the constellation of Gemini.

Discovering nebulae

A few milky blotches can be seen in the sky with the naked eye. With the notable exception of the Magellanic Clouds, which are visible from the southern hemisphere, these are not very remarkable. It was only when an astronomical telescope began to be used that the existence of non-stellar luminous patches could be confirmed. They were first called "nebulae" on account of their diffuse appearance. It was only with the work of Charles Messier and William Herschel, in the 18th century, that these

objects began to be cataloged. The advent of large telescopes, in the 20th century, enabled scientists to discover stars of a very different nature within these blotches.

Planetary nebulae

The Lyra nebula, reference M57 (the 57th object in Charles Messier's catalog), can be seen in the July sky through binoculars, and is a superb specimen of a planetary nebula (see p. 125). It is an immense cloud of expanding gas, mainly composed of hydrogen, helium, carbon and oxygen, which has formed from the remains of an exploded star. The term "planetary" comes from the appearance of this type of nebula: seen through a small instrument most of them have the appearance of clouds of confetti, which in the past reminded people of planets. M27, the Dumbbell Nebula, is another example of a planetary nebula.

Globular clusters

A globular cluster is an extremely concentrated swarm of stars. A cluster like M13, in the constellation of Hercules (see p. 89), contains several hundred thousand stars. In these clusters, stars are condemned to stay linked in this way because of the law of universal gravitation. Such clusters mainly contain old stars and were born at the beginning of the history of the galaxies. Those that we can distinguish are distributed fairly uniformly around our galaxy.

Open clusters

Open clusters, or galactic clusters, are groupings of between several dozen and several hundred young stars, born at the same period, which have not yet had time to disperse into the galaxy. Such groupings are very interesting to astronomers. They enable a comparison to be made between stars of the same age, situated at more or less the same distance from Earth. The apparent differences in magnitude and color are thus only due to the distances between masses. They provide an excellent means of learning about stellar evolution. You can easily see several examples through binoculars, like the Perseus double cluster (see p. 143) or the Pleiades (see p. 178).

Strange nebulosities

Other nebulosities visible in the sky are very spectacular. There are, first of all, the polar aurorae, which can be seen most often in the regions nearest to the poles. But several of these aurorae have recently been seen much further south. They are atmospheric phenomena caused by the arrival of electrically charged particles from the Sun, which then excite the Earth's atmospheric gas.

These occurrences take place every time there is intense solar activity. Due to their fleeting character, they are not considered as being relevant to astronomy.

A polar aurora
This polar dawn was observed in the region of Nice, France, in July 2001.

Supernovae remnants

These are immense clouds of gas produced by an exploding star, and are much more extensive than a planetary nebula. The massive star exploded several thousand years ago, projecting away from it the matter it had retained during the course of its evolution. Such nebulosities can extend over huge areas, which are that much bigger the further back in time the star exploded. The dispersion of matter is such that they are difficult to observe, and they are only revealed through records of observations obtained by large telescopes. The Veil Nebula, in the right wing of Cygnus, the Swan, is a beautiful example that is only accessible to amateurs by means of photographs.

Star nurseries

Some clouds are much easier to observe than supernovae remnants and can even be detected with the naked eye. These are regions of star formation, where there are huge clouds composed mainly of hydrogen and helium. The density of these clouds is very low, just a few atoms per cm^3. They can be seen, despite the predominating empty space, on account of their extraordinary dimensions. These regions are so extensive and so far off that they stand out against their interstellar environment and, illuminated or excited by neighboring stars, become visible to observers on Earth. The Orion Nebula (see p. 214) is the most beautiful example of this type of formation.

Supernova remnants
These residues are found in the Veil Nebula.

Galaxies

Finally, there are nebulae that are accessible to the naked eye. The finest example is the Andromeda galaxy, in the constellation of the same name. A galaxy is an immense cloud of stars. These stars are the biggest in the Universe, grouping together several tens, even several hundreds of billions of stars. Their masses are measured in tens or hundreds of billions of solar masses and their distances are considerable. The Andromeda galaxy, the 31st object in Messier's catalog, is more than two million light years away and contains 200 billion stars. Only two small galaxies are nearer to us than Andromeda, and these are the Magellanic Clouds, two irregular galaxies only visible from the southern hemisphere. The Large Magellanic Cloud is situated 172,000 light years away and the Small Magellanic Cloud is 200,000 light years away. The Magellanic Clouds, the Andromeda Galaxy and a few other galaxies like M33 belong to the same Local Group of galaxies, containing in total about 40 galaxies, of which our own, the Milky Way, is one.

Messier's catalog contains numerous galaxies which, in the 18th century, were impossible to separate from other nebulae. However, progress in observation techniques has since allowed us to do this. Today, thousands of galaxies have been identified, but we know that these space giants are numbered in billions.

How do stars die?

The explosion of a small star
Most of the matter is ejected from the star.

Stars do not go on forever. And their deaths are not peaceful – they explode, sometimes with extraordinary violence. Astronomers are keen to understand these cataclysmic events, even if only to predict the fate of the Sun, which is a star among others, but a star whose survival is of immediate concern to us.

When a star arrives at the end of its life, its core is filled with helium, a residue of the nuclear fusion of hydrogen. This nucleus tends to collapse in on itself and to heat up. The star then becomes unstable and explodes, sending the matter that was part of its long process of evolution into the interstellar environment.

These explosions involve such large quantities of energy that they are easy to observe. But the phenomenon depends largely on the star's mass. The bigger it is, the shorter is its life and the more violent its death.

The explosion of solar-type stars

At the end of its evolution, a star similar in size to the Sun sees its proportional brightness increase considerably, from 10,000 to 1 million times in a few days. Most of the matter is ejected into the interstellar environment, while the center condenses to a small star the size of the Earth, which slowly cools. A large star at the end of its evolution can, in the space of a few days, become as luminous as several billion "normal" stars. This very short-lived phenomenon is called "supernova." The amount of matter released is enormous and the rest of the star contracts into a volume of a few kilometers in radius.

Star explosions are accompanied by an enormous emission of energy in the form of gamma radiation. A system of surveillance enables astronomers to capture "gamma bursts" – very energetic phenomena detected by military satellites. Optical telescopes are automatically programmed to take pictures of the relevant region of the

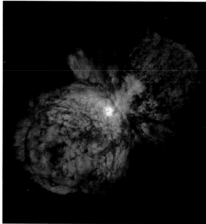

The explosion of a giant
Eta Carinae is one of the most massive stars known at the present time, having no less than 200 solar masses.

sky to try to identify the visible counterpart of the explosion. At present several dozen supernovae a year are detected, mostly in distant galaxies.

The deaths of massive stars: pulsars

When a massive star explodes, its core melts as matter is ejected into the interstellar environment. The remaining mass, which is similar to that of the Sun, is compressed into a sphere of just a few kilometers in radius. The density is enormous, several billion times that of water, and the mass of a thimbleful of this matter would be several billion metric tons. Such a mass is spinning rapidly and, when the magnetic field is intense, it becomes a pulsar whose environment sends out radio waves at the same rhythm as a star rotating. The star then appears to pulsate, hence the name "pulsar" (a contraction of "pulsating star"). Several hundred pulsars are known, and about 50 of these rotate several hundred times per second.

The deaths of very massive stars: black holes

In very massive stars, the concentration of matter is such that even light cannot escape. We are then in the presence of a "black hole." When a black hole appears in a double star, exchanges of matter between the two parts cause an X-ray emission, which can be detected by instruments in orbit above the Earth's atmosphere.

Picture of the Crab Nebula (visible light and X-rays)

Novae and supernovae

In bygone days, when an unknown star appeared in the sky, people greeted this birth by giving it the name of nova, or "new" star. They certainly did not imagine, then, that the star was actually at the end of its life.
Thus, in July 1054, a star was seen to explode in full daylight from both China and America. We now know that this explosion produced a pulsar and a nebula, the Crab Nebula in the constellation of Taurus (M1 in Messier's catalog). In 1572, Tycho Brahe observed a nova in the constellation of Cassiopeia, and Kepler spotted another one 32 years later.

There are about 25 appearances per century of novae that are visible to the naked eye. Out of these, only half a dozen are really spectacular, causing very bright stars to appear. Since 1987, a supernova has been observed in the Large Magellanic Cloud.

The remnants of a nova, NGC 675

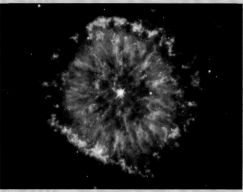

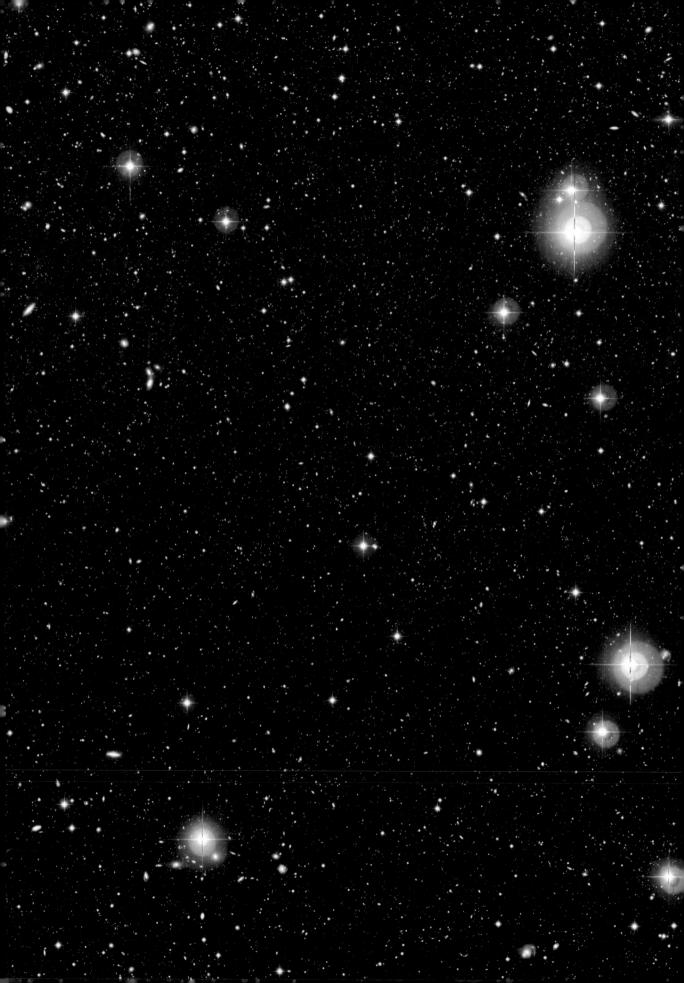

August

Field of galaxies
More than 100,000 galaxies are on this image, which was obtained with the help of the ESO's 2.2 m telescope at La Silla (Chile). Four hundred and fifty photos have been compiled, amounting to more than 50 hours' exposure time.

RENDEZVOUS IN THE SKY

12	Maximum of the Perseids meteor shower
12	The Moon, Venus and Saturn move closer together at the end of the night
18	Maximum of the Kappa Cygnids meteor shower
31	Venus and Saturn move closer together at the end of the night

A month with your head up in the stars

In August, the nights are longer and the summer balminess invites you to observe the sky under excellent conditions. It is the ideal time to look at the Milky Way, with the naked eye of course – but especially through binoculars, which convey an extraordinary impression of depth. Observation of the Perseids, the famous shooting stars of August, will not be hampered this year by the Full Moon. It will be possible to see many meteors, especially at the very end of the night.

In our image
The galaxy NGC 2997, which is about 55 million light years away from us, gives us an idea of the shape of our galaxy, the Milky Way.

Observations of the month

It is at the beginning of these August nights that you should observe the constellation of Lyra, together with Vega, the splendid blue star that is practically at the zenith. Through binoculars, you will discover that the region is much richer than appears at first sight. In good weather conditions, you will even be able to make out the annular nebula of Lyra, an almost star-like object. This planetary nebula is the second to have been discovered: it was seen in January 1779 through an instrument 2.5 inches (63 mm) in diameter.

All times are given in Eastern Daylight Time (EDT). Calculations have been made for a northern latitude of 44° and a western longitude of 80°.

Date	Sun		Moon		Mercury		Venus		Mars		Jupiter		Saturn	
	Rises	Sets	Rises	Sets	Rises	Sets	Rises	Sets	Rises	Sets	Rises	Sets	Rises	Sets
8/1	6:10 a.m.	8:42 p.m.	9:22 p.m.	6:49 a.m.	8:33 a.m.	9:38 p.m.	2:56 a.m.	5:38 p.m.	7:25 a.m.	9:28 p.m.	9:33 a.m.	10:24 p.m.	4:30 a.m.	7:36 p.m.
8/5	6:14 a.m.	8:37 p.m.	11:00 p.m.	11:49 p.m.	8:30 a.m.	9:23 p.m.	2:52 a.m.	5:38 p.m.	7:23 a.m.	9:18 p.m.	9:21 a.m.	10:10 p.m.	4:17 a.m.	7:22 p.m.
8/10	6:20 a.m.	8:30 p.m.	1:02 a.m.	5:17 p.m.	8:17 a.m.	9:01 p.m.	2:49 a.m.	5:38 p.m.	7:20 a.m.	9:06 p.m.	9:07 a.m.	9:52 p.m.	4:00 a.m.	7:05 p.m.
8/15	6:25 a.m.	8:23 p.m.	5:36 a.m.	8:14 p.m.	7:52 a.m.	8:35 p.m.	2:48 a.m.	5:39 p.m.	7:17 a.m.	8:54 p.m.	8:52 a.m.	9:34 p.m.	3:43 a.m.	6:47 p.m.
8/20	6:31 a.m.	8:15 p.m.	11:20 a.m.	10:03 p.m.	7:15 a.m.	8:07 p.m.	2:49 a.m.	5:40 p.m.	7:14 a.m.	8:41 p.m.	8:38 a.m.	9:17 p.m.	3:27 a.m.	6:29 p.m.
8/25	6:37 a.m.	8:07 p.m.	5:34 p.m.	12:42 a.m.	6:31 a.m.	8:33 p.m.	2:51 a.m.	5:40 p.m.	7:11 a.m.	8:28 p.m.	8:24 a.m.	8:59 p.m.	3:10 a.m.	6:12 p.m.
8/30	6:43 a.m.	7:58 p.m.	8:18 p.m.	7:10 a.m.	5:52 a.m.	7:16 p.m.	2:55 a.m.	5:41 p.m.	7:08 a.m.	8:16 p.m.	8:10 a.m.	8:42 p.m.	2:59 a.m.	6:01 p.m.

TIMES OF RISING AND SETTING FOR THE SUN, MOON AND FIVE PLANETS VISIBLE WITH THE NAKED EYE

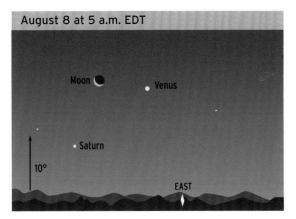

August 8 at 5 a.m. EDT

The planets

Early on the morning of the 12th, just as the night with its abundant shooting stars is ending, the sky will be further embellished by the sight of the crescent Moon, 9° to the east of **Venus** and 13° to the north of **Saturn**. At the end of the night on the 31st, Saturn and Venus move closer together (less than 2° apart) in the constellation of Gemini, also a spectacular sight.

The Perseids take center stage

These meteors can be seen during the whole of the month of August, but their celebrated maximum is on the 12th. They are also known as St. Laurent's tears, since this saint was martyred on August 10, 258. They are actually, more

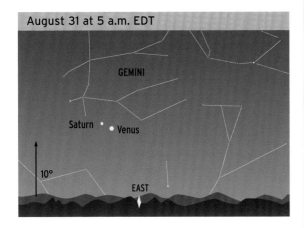

August 31 at 5 a.m. EDT

1 Sunday

2 Monday

3 Tuesday

4 Wednesday

5 Thursday

6 Friday

7 Saturday
- 6 p.m. Last quarter of the Moon
- Maximum of the Southern Iota (ι) Aquarids meteor shower (3 meteors an hour)

8 Sunday

9 Monday

10 Tuesday

11 Wednesday
- 6 a.m. Moon passes the apogee: 405,291 km

12 Thursday
- The Moon, Venus and Saturn move closer together. Visible at the end of the night
- Maximum of the Perseids meteor shower (70 meteors an hour)

13 Friday
- The Moon and Saturn move closer together. Visible at the end of the night (minimum 1.3° at 3:42 a.m.)

14 Saturday

15 Sunday
- 9:24 p.m. New Moon

prosaically, the remnants of the comet Swift-Tuttle coming into annual contact with the Earth. This is a periodic comet which orbits the Sun in 130 years. The meteor shower gets its name from the constellation of Perseus, since the meteors appear to originate from there. It is one of the most spectacular showers, with about 70 meteors an hour moving at a speed of about 60 km/s. It is in the second part of the night that the shooting stars are most numerous, when the visible sky is in the direction of the Earth's movement. The luminous effect that can be seen is simply a trail of hot gas caused by a very small dust particle coming into contact with the Earth's atmosphere. The energy released is such that electrons are pulled from the atoms of the atmospheric gas that it meets. The result is a temporary change in the electrical state of the upper atmosphere, and thus a change in the reflection conditions of the radio waves. This is why shooting stars can be detected in full daylight, with the help of an FM radio receiver. You just need to tune the receiver to a frequency that does not correspond to any broadcast transmission service, so nothing can be heard from the speaker. When a shooting star passes, the conditions for transmitting radio waves undergo a temporary change and a radio station can briefly be heard. Several other meteor showers cross the sky this month. Among them, of course, are the **Delta** (δ) **Aquarids**, which began in July, but also the **Alpha** (α) **Capricornids**, whose maximum is on the 1st, with about 10 meteors an hour. They are reputed to be among the brightest it is possible to see, and they will appear this year on moonless nights.

Shooting stars
Showers of meteorites on summer nights.

A region of the sky to explore: the Milky Way

August is a particularly interesting month for observing our galaxy from the inside. On a dark night it is easy to see a whitish streak across the sky, provided you have some protection from light pollution (see inset p. 145). This is the Milky Way, the trail of Hera's milk that Herakles tried to suck to gain immortality. When the goddess woke with a start, her milk spurted toward the sky, thus causing the Milky Way to appear. It was not until the beginning of the 17th century, thanks to one of the very earliest telescopes, that Galileo showed it was a gigantic cloud of stars. It is possible to make out only the nearest members of this vast family, which includes our own Sun, but it is relatively easy to see outside the galaxy as long as you are careful to look toward the high galactic latitudes – in other words in directions where the Milky Way is less dense, toward the constellation of Ursa Major, for example. Along the bar in the middle of the galaxy, the density of the visible stars increases and marks out the shape of what we call the Milky Way. It is in this milky band that the central region of our galaxy is hidden, in the constellation of Sagittarius. Reach for your binoculars and trace a path along this great circle of the celestial sphere, which goes through the constellation of Cygnus and guides you toward Sagittarius. You will discover some wondrous sights – swarms of stars and clusters of all kinds, and nebulosities inviting you to observe them more closely, with larger telescopes. But don't forget to notice, as the night goes on, how the Milky Way appears to move slightly: as it rotates, the Earth turns toward the outside of the galaxy and reveals to us the vast horizons of the worlds beyond.

Cassiopeia, Andromeda and Perseus

The beautiful constellation of Cassiopeia is at the opposite end of Ursa Major's tail in relation to Polaris. Depending on the observer's position, it has the shape of a W or an M. It never sets and it has a great mythological history. Cassiopeia was empress of Ethiopia, the wife of Cepheus and mother of Princess Andromeda. She claimed that her daughter was more beautiful than the Nereids, the 50 sea nymphs who were the daughters of Nereus. In response to her arrogance, Poseidon sent a sea monster named Cetus (which is where the word "cetacean" comes from) to ravage the kingdom of Ethiopia until such time as Princess Andromeda, who was enslaved, was delivered up to him.

Andromeda was saved by Perseus, who had just performed the feat of

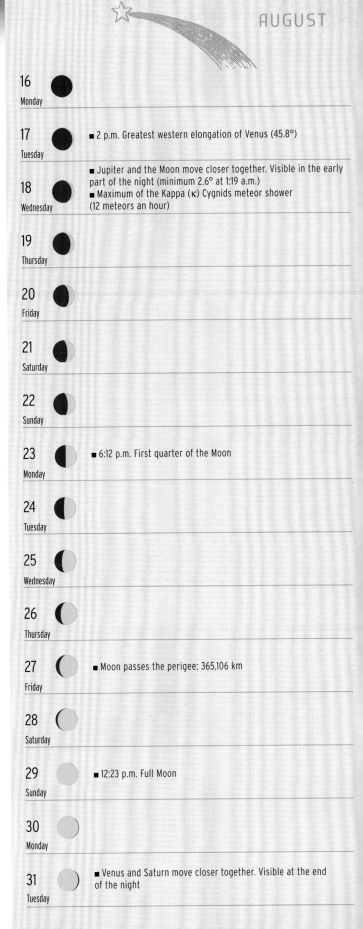

16 Monday

17 Tuesday
■ 2 p.m. Greatest western elongation of Venus (45.8°)

18 Wednesday
■ Jupiter and the Moon move closer together. Visible in the early part of the night (minimum 2.6° at 1:19 a.m.)
■ Maximum of the Kappa (κ) Cygnids meteor shower (12 meteors an hour)

19 Thursday

20 Friday

21 Saturday

22 Sunday

23 Monday
■ 6:12 p.m. First quarter of the Moon

24 Tuesday

25 Wednesday

26 Thursday

27 Friday
■ Moon passes the perigee: 365,106 km

28 Saturday

29 Sunday
■ 12:23 p.m. Full Moon

30 Monday

31 Tuesday
■ Venus and Saturn move closer together. Visible at the end of the night

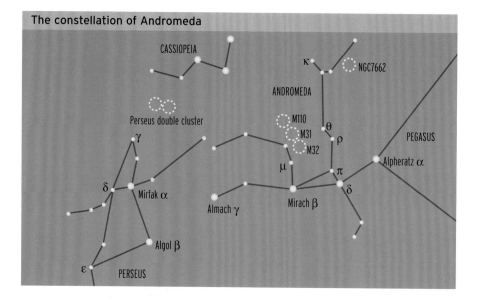

The constellation of Andromeda

decapitating Medusa, the only one of the three Gorgons who was mortal. Medusa's head was a redoubtable weapon; whoever looked at it could not take his eyes away, but was turned to stone. Perseus killed the monster with the aid of this head, saved Andromeda, and married her. In the sky, the Medusa's head is Algol, the name deriving from the Arabic meaning "demon" or "phantom." This star is particularly fascinating since there is a noticeable change in its brightness every three days and it regains its usual radiance within a few hours.

Look at a sky chart and you will see that all the protagonists of this legend are gathered together in the same region.

Alpheratz (or **Sirrah**, or **Alpha** (α) **Andromedae**), from the Arabic *al surat al faras*, "the horse's navel," initially associated with the constellation of Pegasus, represents the princess's head. Situated about 100 light years away, it is a double star whose close companion was detected by spectroscopy: the pair has a period of less than 100 days.

Mirach (or **Beta** (β) **Andromedae**) derives its name probably from *super mirat*, which is found in the Alphonsine tables, or perhaps from the *super mizar* of Ptolemy's Almagest, *mizar* meaning "loincloth." Mirach has a magnitude of 2.03, and is about 75 light years away. It is 75 times more luminous than the Sun. However, you need a good telescope to discern its companion, which has a magnitude of 14 and is situated at 28".

Almach (or **Gamma** (γ) **Andromedae**) is a superb double star discovered by Mayer in 1788. Notice the difference in color of the two components: the main star is orangy yellow, while its companion is blue-green. The principal component itself is a very tightly packed double star, which was detected by spectroscopy. Gamma Andromedae is in fact a quadruple system situated about 260 light years away.

The Andromeda galaxy, M31, is the nearest of the spiral galaxies and the only one visible with the naked eye. M31

THE SKY IN AUGUST

NORTH

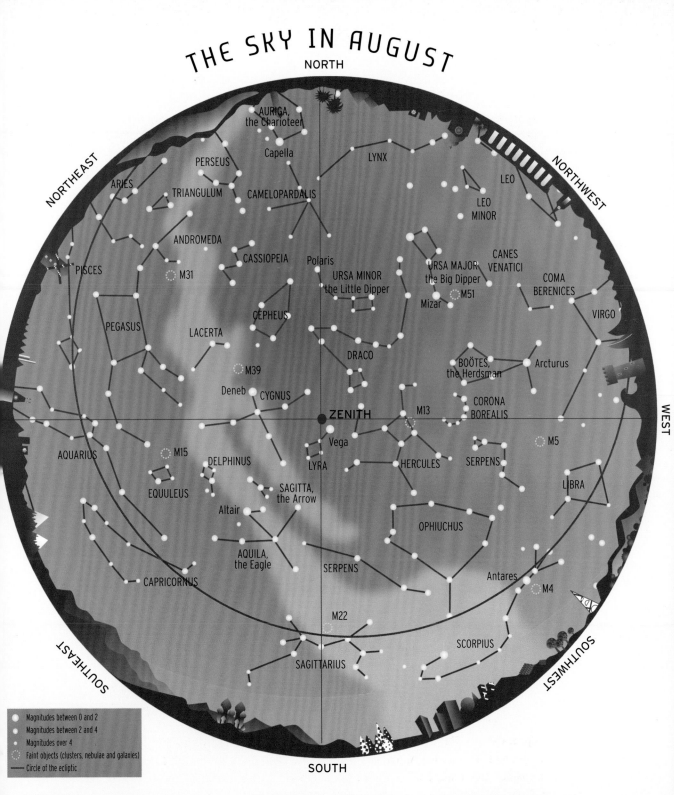

AURIGA, the Charioteer
Capella
LYNX
PERSEUS
LEO
ARIES
TRIANGULUM
LEO MINOR
CAMELOPARDALIS
ANDROMEDA
CANES VENATICI
M31
CASSIOPEIA
Polaris
URSA MAJOR the Big Dipper
COMA BERENICES
PISCES
URSA MINOR the Little Dipper
M51
VIRGO
PEGASUS
CEPHEUS
Mizar
LACERTA
DRACO
BOÖTES, the Herdsman
Arcturus
M39
M13
CORONA BOREALIS
Deneb
CYGNUS
ZENITH
M5
Vega
AQUARIUS
M15
LYRA
HERCULES
SERPENS
DELPHINUS
LIBRA
EQUULEUS
SAGITTA, the Arrow
OPHIUCHUS
Altair
CAPRICORNUS
AQUILA, the Eagle
SERPENS
Antares
M4
M22
SCORPIUS
SAGITTARIUS

NORTHEAST **NORTHWEST**
WEST
SOUTHEAST **SOUTHWEST**

SOUTH

- ○ Magnitudes between 0 and 2
- · Magnitudes between 2 and 4
- · Magnitudes over 4
- ○ Faint objects (clusters, nebulae and galaxies)
- — Circle of the ecliptic

How to use this chart
Hold the chart above your head, matching the word SOUTH that appears at the edge of it with the geographical south of the place you are observing from. Use a compass to help you do this.

This chart shows the sky that is visible at a latitude of 45°
If you are further north or further south, Polaris will be higher or lower in the sky.

Chart of the sky visible at 11 p.m. EDT
at the beginning of the month; at 10 p.m. EDT in the middle of the month; at 9 p.m. EDT at the end of the month.

Barnard 86
This cloud, made up of dark interstellar matter, stands out against the starry background of the Milky Way.

appears to have been noticed in the 10th century, since there is a reference to it in Al Sufi in 905. Simon Marius was the first to observe it with one of the earliest astronomical telescopes, in 1611. M31 is 2.2 million light years away from our own galaxy. Its diameter is 110,000 light years and its mass is on the order of 400 billion solar masses. M31 is surrounded by 140 identified globular masses.

Practice finding the Andromeda galaxy with your naked eye. It is a small, diffuse blotch that will reveal itself further when you use binoculars. A small instrument will give you a magnificent view, but its spiral shape will only become evident through photographs. Take advantage of it to experiment with peripheral vision. You will make out these faintly lit objects much better if you do not look at them directly, so shift your viewpoint by a few degrees in relation to the central line of sight (see p. 144). Very close to M31 you will find **M32** and **M110**, two other galaxies that are accessible to small optical instruments.

NGC 7662 is a superb planetary nebula, also accessible to small instruments. When you first look, you can make out a blurred star with a magnitude of 8, but when you increase the magnification a little, you discover an elliptical structure with the blue-green color characteristic of planetary nebulae.

The Milky Way
This panoramic view was taken from the southern hemisphere.

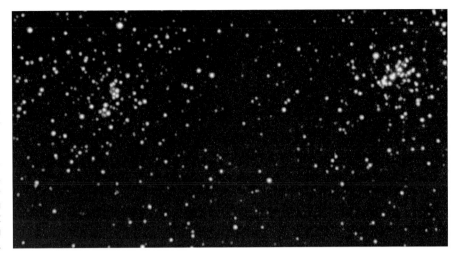

The Perseus double cluster
This double open cluster, visible with the naked eye, is even more splendid when looked at through binoculars.

Mirfak (or **Alpha** (α) **Persei**), "the elbow," is a giant star 4,000 times more luminous than the Sun, situated 570 light years away. Through binoculars, the spectacle is sumptuous: Mirfak is at the center of a very beautiful open cluster 600 light years away.

Algol (or **Beta** (β) **Persei**), "the demon," is the most famous of the eclipsing variable stars. Its variability has been known since antiquity, but it was only established scientifically by Geminiano Montanari (1632–87) in 1667. However, the regularity of its period was demonstrated in 1782 by John Goodricke (1764–86), who put forward the explanation that the partial eclipse was caused by an obscure companion revolving around the principal star. In 1889, the photometric recordings of Hermann Carl Vogel (1841–1907) finally proved the relevance of this theory. The star, which normally has a magnitude of 2.1, changes to a magnitude of 3.4 every 2 days, 20 hours, 48 minutes and 56 seconds. Anybody can observe this phenomenon, with or without instruments.

Algenib (or **Gamma** (γ) **Pegasi**), "the side," is a double star that is difficult to separate (0.4"), with a period of almost 15 years.

The Perseus double cluster, NGC 869 and NGC 884, is one of the wonders of the sky. With the naked eye, two diffuse patches can be distinguished. Now look at them through binoculars. A Chinese legend says that they are the heads of the astronomers Hsi and Ho, who worked at the court of the Emperor Tsung-k'ang in the 3rd millennium B.C. Their role consisted mainly in predicting eclipses, which at that time it was believed were caused by a dragon devouring either the Sun or the Moon. When an unpredicted eclipse occurred, the drunken astronomers were accused of not having taken the necessary steps to chase away the dragon. They were therefore decapitated, and their heads are always to be seen in the sky, reminding astronomers that they must complete the tasks they have been entrusted with.

Practical astronomy

Knowing how to observe with the naked eye

Until the beginning of the 17th century, this was the only possible means of observation. The shape and size of the Earth, the distance separating the Earth from the Moon and the laws governing the movement of the planets were all found without any form of telescope. Contrary to widespread belief, it is quite possible to practice astronomy without any optical instrument. So let us learn how to use our eyes again to explore our environment

An exceptional instrument

The eye is an organ of extraordinary precision that we generally use rather badly, especially for night observation. It is in fact capable of seeing very well at night, but it needs quite a long time to adapt. When first adapting to the dark, the pupil opens, and its diameter may go from 2 to 8 millimeters. When it is fully open, this allows about 16 times more light to pass into it than in full daylight. This opening of the pupil is not the only way the eye adapts to the dark. The sensitive surface of the eye, the retina, is lined with two types of receptor cells. The cones, which are color-sensitive, adapt very quickly but very poorly to low light. The rods adapt slowly to low light, but are not very color-sensitive. After about half an hour, the eye is capable of detecting lights a billion times fainter than it registers in full daylight. It is therefore essential to be patient and to train your eye for night vision if you wish to take full advantage of it.

Peripheral vision

Another problem with night viewing comes from the disparities in the retina's sensitivity. The cones are gathered near to the eye's axis, and the rods are around this central area. The area that is most sensitive to low light is slightly displaced in relation to the center of the retina. The eye therefore has to look slightly askance in order to distinguish the most obscure objects. This is known as using "peripheral vision."

The human eye's ability to separate is very variable, and this is primarily what ophthalmologists measure when carrying out eye tests. It is considered that a "normal" eye separates 1' of arc, or 1 millimeter at 3.4 meters. If this is the case, it receives a grade of 10. Some people have a greater ability to separate, which explains why it is possible to have grades of 11/10, or even 13/10.

You can test your visual acuteness by observing the second star in the Great Bear's tail, in which a good eye should be able to discern the two components, Mizar and Alcor, without any difficulty. Some people can spot Jupiter's four main satellites, or at least the largest among them, without an instrument. Another test of visual acuteness can be carried out on the Pleiades cluster: while five stars in it can easily be identified, some people can clearly distinguish seven or even nine.

A few tricks

When they are carrying out observations, astronomers avoid being dazzled

Light pollution seen from space
This picture was produced from observations carried out for the DMSP (Defense Meteorological Satellite Program). It shows the invasion of artificial light in the northern hemisphere.

by using tricks that you can easily adopt yourself. When you have to read documents or enter a brightly lit building, try to "save" one eye, either by keeping it closed or by covering it with an eye patch. In this way you will keep it accustomed to darkness, and you will be able to resume your observations very quickly.

Another tip, if you have to read a sky chart, is to use a red light, which is much less dazzling than white light. The usual trick is to coat the bulb of your "astronomical" torch with nail polish!

Light pollution

The need to adapt to darkness reveals one of the plagues of our time: light pollution. Our cities are extremely brightly lit, and so we send an enormous quantity of light into the sky.

This light illuminates the particles suspended in the atmosphere and masks the stars. Away from cities, the sky is much darker, but it is becoming difficult, even in the countryside, to find a site that is not affected by a street lamp or some other light source. For professionals, the problem is even greater. In the northern hemisphere, there is no longer any site that is not polluted by light to some degree.

All those wishing to observe the sky should act, at a local level, to ensure that public lighting is directed toward the ground and not toward the sky. That way the community will save both energy and money, and the panorama of the sky will continue to be accessible to enthusiasts of the cosmos.

Introduction to astronomy

Discovering the galaxies

A double galaxy
NGC 2207 and NGC 2163 are two galaxies interacting with one another in the constellation of Canis Major.

The Milky Way is what we can see from our own galaxy. In Classical times, this stream of light across the sky was considered to be a circle drawn on the celestial sphere. Thought to be the trail of Hera's milk which had shot into the sky due to the eagerness of the infant Herakles, this circle was called *galaxya* (from the Greek *gala*, "milk," in the phrase *galaxias kiklos*, the "circle of milk"). The term "galaxy" was born out of this legend, and was used thereafter to refer to clouds of stars. Before we could understand the nature of our galaxy, it was necessary to discover other galaxies: since we are condemned to remain within our cloud of stars, it is impossible for us to see its shape and dimensions. However, from simple observations, we can obtain some idea of the structure of the system in which is found our own Sun. It is clear that the stars are not distributed uniformly around the Sun and that their density increases when we look in the direction of the Milky Way. This being the case, we can then conclude that they are spread out over an immense disk.

The geography of our galaxy

Our galaxy is a gigantic complex containing about 100 billion stars. The diameter of the Milky Way disk is about 100,000 light years, and it is some 2,000 light years thick. The Sun is a little less than 30,000 light years from the center of the system, around which it turns at 220km/s. The age of the Milky Way is estimated to be 12 billion years. The galaxy is surrounded by a vast system of globular clusters spread over an enormous sphere 150,000 light years in diameter.

The hot young stars are positioned along the arms of the galaxy, near the place where they were formed. The old stars are situated more toward the center and in the globular clusters.

Our galaxy is said to be "spiral." The arms of the spiral situated near to the Sun are those of Sagittarius, Orion and Perseus, and they are characterized by open clusters and regions of star formation. The Sun, which has turned around the galaxy about 20 times since its birth, goes from one arm to another in a few tens of millions of years. Its environment is made up of a group of about 20,000 stars, called "local" stars, attached to the arm of Orion and extending over approximately 2,000 light years.

The various galaxies

We now know several thousand galaxies, which began to be studied systematically when large telescopes came into use in the 1920s. Like stars, galaxies are classified according to their characteristics. But unlike the former, different shapes can be distinguished, the chief ones being elliptical, spiral and irregular.

– Elliptical galaxies only contain stars. They are shaped like an ellipsoid but

The Sombrero galaxy
The hub of this galaxy reveals intense activity.

At the heart of the galaxies

The density of matter increases in the central regions (hubs) of the galaxies. However, the light observed in these regions does not resemble the light emitted by a star cluster. Analysis of this radiance has led to the discovery that it is not simply thermal, produced by heated gas, but that a large part of it comes from the movement of electrons, which reach speeds close to the speed of light and are then slowed down by an intense magnetic field. The hub of a galaxy is not, then, simply a concentration of stars.

In spiral galaxies like our own the hub shelters an expanding disk of gas, which is the center of intense star formation. Nearer the center still, the temperature increases to around 10,000 K. The area immediately around the center is complex and contains a strong radio source confined within an extremely restricted area of only a few billion kilometers. It is now thought that the center of our galaxy is thus composed of a "tepid" disk of gas spread out between 5 and 30 light years. Inside this disk nestles a very dense star cluster whose total mass is almost two million solar masses. At the heart of this cluster is a black hole of five million solar masses of fairly restricted volume, i.e., a diameter of only about 30 million kilometers. The stars revolving in this region do so at great speed, and one of them is occasionally hurled toward the black hole. The forces at play may then deform and even destroy it, causing fits of luminous activity that are visible for several decades. It is thought that our galaxy suffers this kind of accident once every 10,000 years, but observations confirm that the hubs of distant galaxies are much more frequently at the center of this kind of cataclysm.

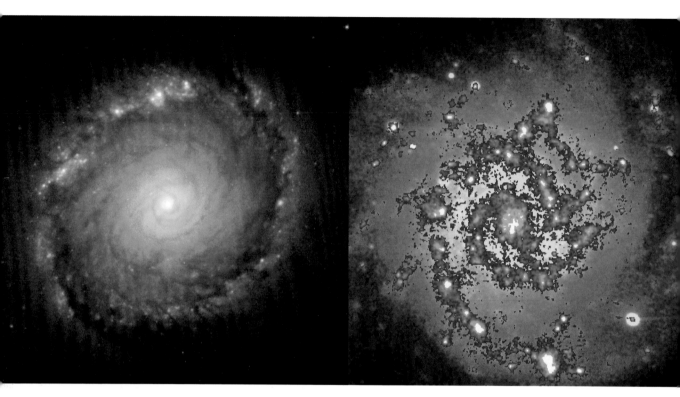

NGC 1512
This picture was obtained by combining seven photos taken by the Hubble Space Telescope.

are of very variable dimensions, some containing only a few million stars and others several hundred billion. Apparently the lack of gas prohibits star formation here.

– Spiral galaxies feature two or four spiral arms attached to a central hub. They contain dust, gases and stars of all ages. These spiral arms originate when a density wave passes, compressing the matter slightly and causing stars to appear. Massive stars, which evolve very rapidly, stay near to the place where they were formed, whereas small stars go round the galaxy several times, even several dozen times, during the course of their evolution.

– Irregular galaxies are often of small dimensions, relatively speaking, and their shape has not been molded by strong gravitation.

Clusters of galaxies

Galaxies are grouped in clusters with anything from a few dozen to a few hundred members. Several thousand clusters are now known. As with stars, the studies that have been made of them enable us to begin to imagine their evolutionary pattern.

When we observe the most distant clusters, we are seeing them as they were several billion years ago. The galaxies they contain were then young. However, these distant clusters are richer in spiral galaxies than clusters that are closer to us, and which we are seeing with less of a time-lag. It seems, therefore, that the giant elliptical galaxies common in the nearer clusters were formed as they fed progressively on the spiral galaxies that passed close to them.

M74
This picture, processed by the Hubble Space Telescope, reveals the distribution of hot young stars in the spiral arms of the galaxy.

NGC 3314
These two galaxies are situated in the same line of sight, but are not interacting with one another.

It was studying distant galaxies that enabled Edwin Hubble to reveal the expansion of the Universe (see p. 75). Recession velocity can be directly observed thanks to the Doppler–Fizeau effect, which shifts the spectral lines according to the source velocity. When the source moves nearer, the shift is toward the shorter wavelengths, i.e., toward the blue. When it moves away, the shift is toward the longer wavelengths, i.e., toward the red.

The Milky Way belongs to a small group of neighboring galaxies, among which are the Magellanic Clouds, the Andromeda galaxy, M33, and a few galaxies which can be seen in the constellations Leo, Sculptor, Ursa Minor and Draco. This group is part of a "supercluster" called the Local Super-cluster, which is centered on the Virgo cluster and extends over a diameter of 100 million light years. The mass of the Local Supercluster is around a million billion solar masses. The Local Group of galaxies, to which we belong, is moving within this at a rate of 600 km/s toward the Hydra–Centaurus Supercluster and the Virgo Cluster. All these groups together are pulled by a vast con-centration of galaxies, called the Great Attractor, situated 200 million light years away and with a mass 200,000 times greater than that of our own galaxy.

NGC 4622
This superb spiral, despite appearances, is revolving clockwise. Hot young stars are clearly visible in the outer arms.

New observation tools

A telescope gathers light from the stars by means of a mirror: the larger its surface area, the more luminous the image (thus allowing stars that are further away to be observed). In the same way, fineness of detail or "resolution" depends on the diameter of the mirror. Astronomers are therefore perfecting larger and larger telescopes in order to probe the Universe. The mirror at Mount Palomar in California, which had a 16.4-foot (5 m) opening, was for a long time the foremost example. Then, the record was held by the two telescopes at Keck in Hawaii, which have two mirrors each measuring 31.2 feet (9.8 m) in diameter. But in 2003, the Large Telescope in the Canary Islands, with its 34.1-foot (10.4 m) opening, took the number one spot. The manufacture of giant mirrors is a job for experts: their curve has to be exact to the nearest millimeter for the image to be clear. But the bigger the mirror, the more it becomes bent under its own weight. Engineers have found ways of solving this problem: instead of casting a giant mirror in a single piece, they create a mosaic of small mirrors that they join together. Thus, the Keck telescope is made up of 36 hexagons with sides measuring 90 centimeters.

Another solution is not to take the mirror's rigidity into consideration but to continually correct its deformities by means of computer-assisted hydraulic jacks (the so-called "adaptive" approach). At the same time, it is possible to erase any distortions of image arising from atmospheric turbulence. This is known as the "active" approach, and gives excellent results.

Addition of images

Another promising method comes from the use of interferometers. The laws of optics teach us that by superimposing two images from two separate telescopes, luminosity can be obtained which is equal to that of a virtual mirror with the diameter of both mirrors added together. As for the resolution of the image, this is no longer dependent on the diameter of the mirrors, but on the distance separating them, which may reach hundreds of meters. However, the images need to be superimposed in stages, since the beams of light from each mirror have to cover the same distance, to within a fraction of a micron, before they can be combined.

This method using interferometers is in the process of being tested at the ESO

M16
An example of a picture obtained by the Hubble Space Telescope in a region of star formation.

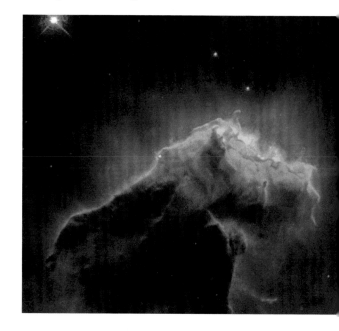

Hubble's successor

The Hubble Space Telescope was put into orbit in April 1990 by the US Space Shuttle. It includes a mirror 7.9 feet (2.4 m) in diameter, which focuses starlight onto a battery of interchangeable instruments: movie camera, infrared spectrometer and ultraviolet spectrograph. It is maintained through flights by the Shuttle: thus, its badly designed mirror (it was short-sighted) was repaired in orbit by astronauts three times in 1993. Three other interventions took place in 1997, 1999 and 2001 to replace defective gyroscopes and install new instruments.

Freed from atmospheric disturbances and composed from different forms of radiation, the beautiful pictures taken by the Hubble Telescope have been all round the world. But NASA and ESA are already drawing up plans for its successor, the Next Generation Space Telescope or NGST. It will be placed 1.5 kilometers from the Sun so that it does not suffer from the parasitical radiation from our planet. The NGST will take with it a folded mirror, which will open up in orbit to reach 7 meters in diameter. It will observe the Universe in infrared. Thanks to the exceptional luminosity of its mirror, it will be able to collect pictures from the most distant galaxies, enabling us to go back in time and observe the Universe as it was at its very beginning.

The VLT
The four telescopes of the VLT (Very Large Telescope) each have a mirror 8.2 m in diameter.

(European Southern Observatory) in Chile. At the beginning of 2004, the images from its four 26.9-foot (8.2 m) diameter telescopes will converge – by means of an interplay of mirrors and underground tunnels – in an optical chamber halfway between the domes. This group, known as the VLT (Very Large Telescope) will then have the resolution of a virtual telescope 656.2 feet (200 m) in diameter.

Telescopes in orbit are also destined for a brilliant future, since they are free of the turbulence contained in the Earth's atmosphere and can therefore collect perfectly clear images. Some of these, like the Hubble Space Telescope, operate in visible light as well as in ultraviolet and infrared – these wavelengths are inaccessible from Earth because they are blocked by the Earth's atmosphere. Other satellites are gathering X-rays and gamma rays, emitted by violent incidents in the cosmos, like stars exploding and matter being sucked into black holes (see p. 133).

The NGST

September

Moonrise
The Moon emerging from the Earth's atmosphere, as seen from the Columbia space shuttle in January 2003.

A month with your head up in the stars

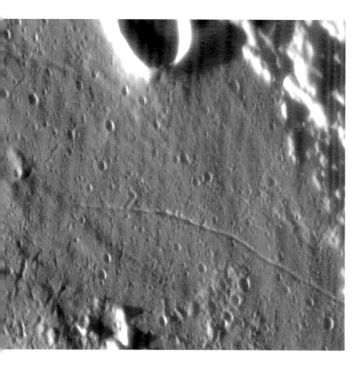

The month begins with the Moon lighting up the skies in the early evening. Mars is still clearly visible on the southern horizon as soon as the Sun has set. September is the month of the autumn equinox: the nights are now getting longer, which means that the sky can be viewed in the evenings, at very convenient hours.

Sunset on the crater Taruntius
The resolution of this image of the Moon taken by the VLT is of the order of 130 m, a record of precision.

Observations of the month

The autumn equinox is approaching. The direction of the sunrise is moving toward the east. From the 23rd, the Sun enters the southern hemisphere, and the nights in the northern hemisphere will now be longer than the days. If you make daily observations of the directions in which the Sun rises and sets, you will be able to appreciate the slow shift drawing them toward the south. Arcturus is beginning to disappear below the northwest horizon, while Capella shines in the northeast, heralding the imminent arrival of Aldebaran and the stars of Taurus toward the east. The constellation of Pisces is ascending in the sky in

All times are given in Eastern Daylight Time (EDT). Calculations have been made for a northern latitude of 44° and a western longitude of 80°

TIMES OF RISING AND SETTING FOR THE SUN, MOON AND FIVE PLANETS VISIBLE WITH THE NAKED EYE														
Date	Sun		Moon		Mercury		Venus		Mars		Jupiter		Saturn	
	Rises	Sets	Rises	Sets	Rises	Sets	Rises	Sets	Rises	Sets	Rises	Sets	Rises	Sets
9/1	6:45 a.m.	7:55 p.m.	9:02 p.m.	9:29 a.m.	5:39 a.m.	7:10 p.m.	2:56 a.m.	5:41 p.m.	7:07 a.m.	8:11 p.m.	8:04 a.m.	8:35 p.m.	2:46 a.m.	5:47 p.m.
9/5	6:49 a.m.	7:46 p.m.	3:40 p.m.	2:03 a.m.	5:23 a.m.	7:03 p.m.	3:01 a.m.	5:40 p.m.	7:05 a.m.	8:01 p.m.	7:53 a.m.	8:21 p.m.	2:32 a.m.	5:32 p.m.
9/10	6:55 a.m.	7:37 p.m.	2:16 a.m.	6:17 p.m.	5:20 a.m.	7:00 p.m.	3:07 a.m.	5:40 p.m.	7:02 a.m.	7:45 p.m.	7:39 a.m.	8:04 p.m.	2:15 a.m.	5:14 p.m.
9/15	7:01 a.m.	7:28 p.m.	7:56 a.m.	8:09 p.m.	5:33 a.m.	7:00 p.m.	3:14 a.m.	5:38 p.m.	6:59 a.m.	7:32 p.m.	7:24 a.m.	7:43 p.m.	1:58 a.m.	4:56 p.m.
9/20	7:06 a.m.	7:18 p.m.	2:15 p.m.	9:53 p.m.	5:57 a.m.	7:01 p.m.	3:22 a.m.	5:36 p.m.	6:56 a.m.	7:19 p.m.	7:10 a.m.	7:25 p.m.	1:40 a.m.	4:38 p.m.
9/25	7:12 a.m.	7:09 p.m.	6:20 p.m.	3:22 a.m.	6:25 a.m.	7:01 p.m.	3:31 a.m.	5:33 p.m.	6:53 a.m.	7:06 p.m.	6:56 a.m.	7:08 p.m.	1:22 a.m.	4:20 p.m.
9/30	7:18 a.m.	7:00 p.m.	7:44 p.m.	9:29 a.m.	6:53 a.m.	7:00 p.m.	3:41 a.m.	5:29 p.m.	6:50 a.m.	6:53 p.m.	6:42 a.m.	6:50 p.m.	1:04 a.m.	4:01 p.m.

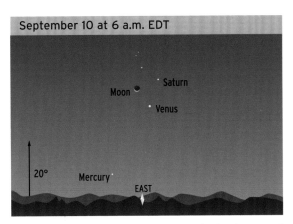

September 10 at 6 a.m. EDT

Saturn
Moon
Venus

20°   Mercury
EAST

the early evening beneath the Square of Pegasus, while Ursa Major is descending toward the north.

Among the curiosities observable with the naked eye, note the variations in brightness of Algol, a star situated in the constellation of Perseus. Algol is an eclipsing variable star: in fact, it is a system of two stars revolving one around the other with a very short period of a little under three days. Thus a large, not very bright star, regularly masks a smaller, brighter star, to the extent that the variations in brightness are perceptible to the naked eye. The minima of Algol which are observable under good conditions are listed each month in the columns headed *The sky from day to day* (right).

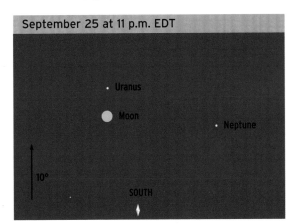

September 25 at 11 p.m. EDT

· Uranus

○ Moon

· Neptune

10°

SOUTH

1 Wednesday		
2 Thursday		
3 Friday		
4 Saturday		
5 Sunday		
6 Monday	■ 11:11 a.m. Last quarter of the Moon
7 Tuesday	■ 11 p.m. Moon passes apogee: 404,464 km
8 Wednesday	
9 Thursday	
10 Friday	■ The Moon, Venus and Saturn move closer together. Visible at the end of the night
11 Saturday	
12 Sunday	■ The Moon and Mercury move closer together. Visible at the end of the night (minimum 3.7° at 8:56 p.m.)
13 Monday	
14 Tuesday	■ 10:30 p.m. New Moon
15 Wednesday	

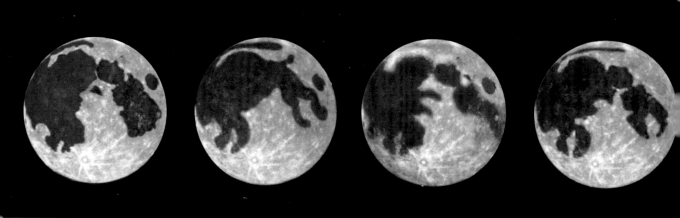

The shapes of the dark patches on the Moon
The dark patches visible on the Moon have been popularly described as shaped like a woman's head, a man's head, a rabbit or a crab.

The planets

You can make excellent observations with the naked eye at the tail end of the night, thanks to the relative proximity of **Venus** and **Saturn** between which, early on the morning of the 10th, you can see the crescent Moon. This month, Venus is crossing the constellation of Cancer, leaving Saturn and Gemini on the 4th to reach Leo on the 24th. At 11 p.m., the Moon is 4° beneath **Uranus**, practically full south. It is a good opportunity to discover this planet, which is invisible to the naked eye but accessible with binoculars or small telescopes. On the 29th, around 3 a.m. the planets **Mercury, Mars** and **Jupiter** will be gathered together in a field of less than 1°. Since the Sun is 5° away from this conjunction of the planets, it would be a mistake to try to observe it with the naked eye. However, it will be accessible thanks to the SOHO satellite, and can be viewed on the Internet.

Visit SOHO's site of wide-angle corono-graph images: http://sohowww.estec.esa.nl/data/realtime-images.html, from the 20th, and you will see a magnificent sight.

The Moon takes center stage

At the time of the new Moon, the sky is graced with a very thin crescent that is sometimes difficult to see. This beautiful early morning sight should be viewed from the 9th onward. It will be doubly interesting, as the Moon will then have taken its place in a procession of planets made up of Saturn, Mercury and Venus. Take the opportunity this month to measure the Moon's movement: you will be able to see it in the constellation of Pisces on the 1st and 2nd, then it will cross Aries on the 3rd and 4th, Taurus from the 5th to the 7th, Gemini on the 8th and 9th, and finally Cancer from the 10th, before disappearing in the light of the Sun in Leo. The Moon moves relative to the stars by about a dozen degrees each day, which corresponds to the part of the sky hidden by your hand when you stretch out your arm.

The Moon is the object above all other that is constantly changing. If these

DEFINITION ▶

Earthlight or earthshine: this is light from Earth reflected onto the Moon. When, from Earth, we can only see a thin crescent, the Sun is mainly illuminating the hidden face of the Moon. Since the Sun is in the same half of the sky as the Moon, the Earth is illuminated directly. By means of reflection, the Earth illuminates the sky, lighting up the Moon's surface just as the full Moon lights up the surface of the Earth. Earthlight can be seen every time there is a thin crescent Moon.

▶ DEFINITION

Gibbous: this word comes from the Latin gibbosus, "hunch-backed," and means that more than half of the Moon is lit up. The Moon is gibbous between the first quarter and the full Moon, and between the full Moon and the last quarter.

apparent changes are now-adays well understood, in the past they troubled observers for a long time, and it was very difficult to develop mathematical models allowing the posi-tion of the Moon to be accurately calculated. Now-adays, the Moon's position is known to within a few centimeters. The main characteristic of the Moon is its changing aspect, and this is what makes it a natural calendar: everyone can see the Moon, and can therefore use its appearance as a time indicator. As the same phases recur on average every 29.53 days, the grouping together of days in thirties was natural, and this was the origin of our months. These phases follow one another thus: thin crescent, first quarter, gibbous ◗ Moon becoming bigger by the day, full Moon at the moment it is passing opposite the Sun, gibbous Moon, last quarter, crescent growing smaller until it becomes the new Moon. Recognizing the phase is very easy: you just need to remember that the Moon that is visible in the evening is near

The phases of the Moon
The relative positions of the Sun-Earth-Moon trio determine the system of the Moon's phases.

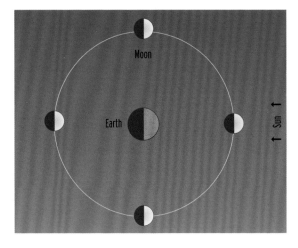

16 Thursday	
17 Friday	
18 Saturday	
19 Sunday	
20 Monday	
21 Tuesday	■ 11:54 a.m. First quarter of the Moon
22 Wednesday	■ 12:31 p.m. Autumn equinox ■ 5 p.m. Moon passes the perigee: 369,600 km
23 Thursday	
24 Friday	
25 Saturday	■ The Moon and Uranus move closer together. Visible in the middle of the night (minimum 4° at 11 p.m.)
26 Sunday	
27 Monday	
28 Tuesday	■ 9:10 a.m. Full Moon
29 Wednesday	■ Mercury, Mars and Jupiter move closer together. Visible at the end of the night on the Internet, via SOHO
30 Thursday	

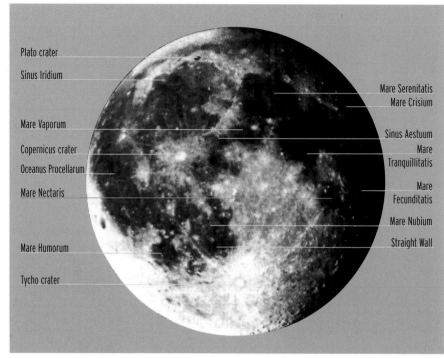

Plato crater
Sinus Iridium
Mare Vaporum
Copernicus crater
Oceanus Procellarum
Mare Nectaris
Mare Humorum
Tycho crater

Mare Serenitatis
Mare Crisium
Sinus Aestuum
Mare Tranquillitatis
Mare Fecunditatis
Mare Nubium
Straight Wall

The main feartures to be seen on the Moon
The Moon is shown here with the north at the top, as in a direct view. If you use an instrument that reverses the image, adjust your markers in order to identify the features.

to the first quarter and the Moon visible in the morning is near to the last quarter. There is one apparent movement of the Moon that is easy to observe, and that is its dance around the ecliptic line. Each month, the Moon ascends and descends by a little over 5°, about 10 times its diameter, on either side of the Sun's apparent trajectory. Each occasion that the Moon passes over the ecliptic corresponds to what is called a node of the orbit. Every lunar month, the Moon passes through two nodes of its orbit.

The Moon provides an excellent means for learning about observation. Try to practice drawing what you can see with your naked eye, through binoculars or with a telescope. You will be surprised at the amount of detail revealed by close observation.

Use the adjacent chart to identify the principal features of the Moon. With the

naked eye, the main seas look like dark and relatively uniform patches. The Mare Crisium, or Sea of Crises, which is vast and circular, 370 kilometers in diameter, can be observed from the first days of the

A few observations to carry out

Age of the moon	Bright objects of interest
2 days	Mare Crisium; Langrenus, Condorcet, Jansen and Gauss craters
3 days	Mare Fecunditatis; Cleomedes and Messier craters
4 days	Mare Nectaris; Proclus and Frascator craters
5 days	Mare Tranquillitatis; Plana, Capella and Fabricius craters
6 days	Cyrillus, Catherine, Cassini and Theophilus craters
7 days	Mare Vaporum; Ptolemaeus and Hyginus craters; Apennine mountains
8 days	Mare Frigoris; Archimedes and Plato craters; the Straight Wall
9 days	Craters Tycho, Clavius and Eratosthenes
10 days	Mare Nubium; Sinus Iridium; craters Copernicus and Gassendi
11 days	Craters Campanus and Mercator
12 days	Mare Imbrium; crater Aristarchus
13 days	Craters Grimaldi and Hevelius
14 days	Craters Riccioli and Pythagorus
On the 14th day, the Moon is full, and then the structures gradually disappear again.	

THE SKY IN SEPTEMBER

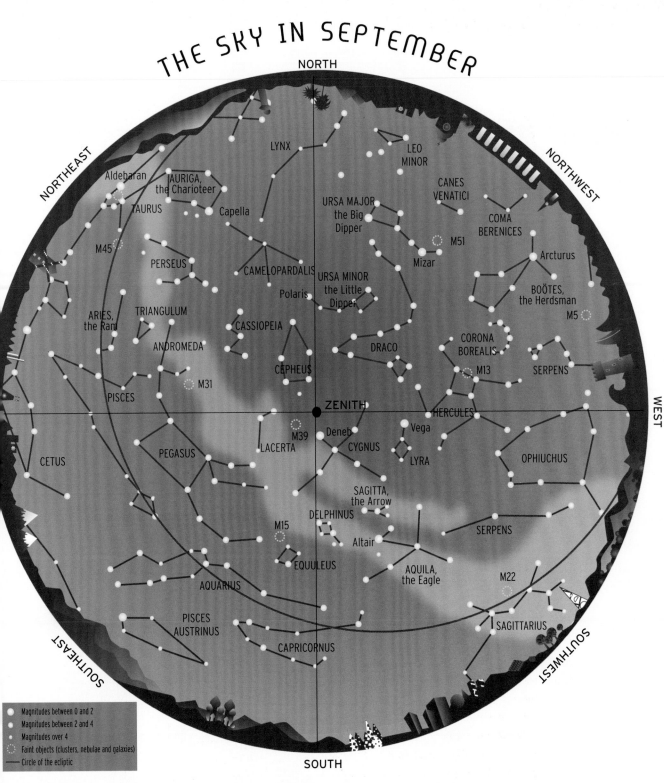

NORTH

LYNX

LEO MINOR

NORTHEAST

NORTHWEST

Aldebaran

AURIGA, the Charioteer

CANES VENATICI

TAURUS

Capella

COMA BERENICES

URSA MAJOR the Big Dipper

M51

M45

Arcturus

PERSEUS

CAMELOPARDALIS

Mizar

URSA MINOR the Little Dipper

BOÖTES, the Herdsman

Polaris

DRACO

M5

ARIES, the Ram

TRIANGULUM

CASSIOPEIA

CORONA BOREALIS

ANDROMEDA

M13

SERPENS

CEPHEUS

M31

HERCULES

PISCES

ZENITH

WEST

M39

Deneb

Vega

CETUS

LACERTA

CYGNUS

OPHIUCHUS

PEGASUS

LYRA

SAGITTA, the Arrow

DELPHINUS

SERPENS

M15

Altair

EQUULEUS

M22

AQUILA, the Eagle

AQUARIUS

SAGITTARIUS

PISCES AUSTRINUS

CAPRICORNUS

SOUTHEAST

SOUTHWEST

- ● Magnitudes between 0 and 2
- ● Magnitudes between 2 and 4
- ● Magnitudes over 4
- ◌ Faint objects (clusters, nebulae and galaxies)
- — Circle of the ecliptic

SOUTH

■ **How to use this chart**
Hold the chart above your head, matching the word SOUTH that appears at the edge of it with the geographical south of the place you are observing from. Use a compass to help you do this.

■ **This chart shows the sky that is visible at a latitude of 45°**
If you are further north or further south, Polaris will be higher or lower in the sky.

■ **Chart of the sky visible at 11 p.m. EDT** at the beginning of the month; at 10 p.m. EDT in the middle of the month; at 9 p.m. EDT at the end of the month.

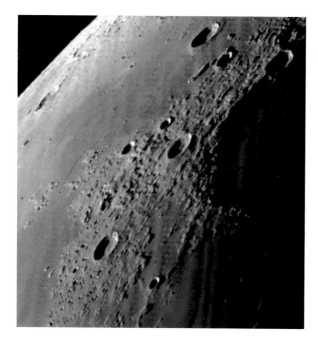

Lunar landscape
To the northwest of the Moon you can see the Sinus Roris, with the Marius and Sharp craters.

A simple pair of binoculars gives a completely different dimension to the Moon. The ideal moments to choose are those when the light comes from a low angle, so that the shadows are clearly shown. Thus, it is not at the time of the full Moon that observations are the most interesting and spectacular. Following the example of Edmund Halley, one should observe the Moon daily in order to reap full benefit from the wonders it has to offer, and get to know it intimately. From the sixth day on, try to practice finding the craters Theophilus, Cyrillus, Cassini and Catherine; on the ninth day, Tycho; and on the tenth, Gassendi and Copernicus, easily recognizable in the Ocean of Storms (Oceanus Procellarum), at the center of the bright rays around it. On the 12th day, you can observe the craters Aristarchus and Hipparchus. When using either a telescope or a spyglass, choose what you want to observe according to an object's degree of illumination (and therefore according to the age of the Moon).

lunar month. In the first quarter, the great Seas of Fecundity (Mare Fecunditatis), Tranquillity (Mare Tranquillitatis) and Serenity (Mare Serenitatis) are lit up. It is from the full Moon onward that the Seas of Clouds (Mare Nubium), Rain (Mare Imbrium) and Moisture (Mare Humorum) are visible.

A region of the sky to explore: Pisces

Pisces is an inconspicuous constellation, whose stars have a brightness of magnitude 4 or less. The celestial equator and the ecliptic cross one another in the constellation, and the Sun also passes through it at the spring equinox; Pisces therefore rises due east and sets due west. It is a constellation that has been known about since the time of the earliest civilizations in Mesopotamia, and is made up of a pair of fish, one in

the north, to the east of the square of Pegasus, and the other in the south. A Greek legend recounts how Aphrodite and her son Eros, frightened by the monster Typhon while bathing, metamorphosed into fish in order to escape him. They were careful to bind themselves to one another so as not to be separated. The fish were put into the sky in memory of this incident. For the Romans, the protagonists were known as

Venus and Cupid, with Typhon keeping his own name. According to some sources, the episode is supposed to have taken place on the banks of the Euphrates, in Syria, and this is why the Syrians gave up eating fish: they feared that, while fishing, they might attack the refuge of the gods, and that they might accidentally devour Venus and Cupid while eating.

Alrisha (or **Alrescha** or **Alpha** (α) **Piscium**) comes from the Arabic *al rischa*, "the rope." The star represents the knot tying the ropes to which the two fishes are attached. This double star was discovered by William Herschel in 1779. Since then, it has become more difficult to identify, because the separation is constantly diminishing (at present it stands at 1.9"). The system's period is 720 years, and the periastron (the point when the two stars will be closest to one another) will be reached in 2060. Each of the two components is itself a binary, so we are seeing here one of the numerous quadruple systems that populate our galaxy. This particular system is about 130 light years away.

19 Piscium, a faint star that is nevertheless accessible to small instruments, is characterized by its strange red color that can be seen with a simple pair of binoculars. It is a good test for determining color. This star is a cold giant situated about 1,000 light years away.

Wolf 28, van Maanen's star, is one of the rare white dwarfs that can be seen with a small instrument. It is 2° to the south of Delta (δ) Piscium, the sixth star from Alpha (α) Piscium when you follow the line leading to the southernmost fish. Its apparent magnitude is 12.4, but

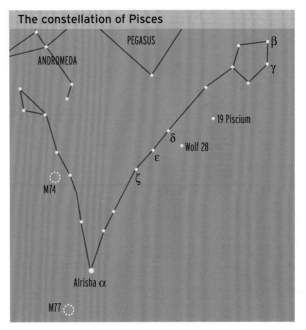

The constellation of Pisces

it is very near to the Sun at 13.8 light years away. Its intrinsic luminosity is 6,000 times less than the Sun's, and its surface temperature reaches 6,000 K. It is only 12,500 kilometers in diameter – about the same as the Earth's diameter. With a mass comparable to that of the Sun, this star has a density of one million (the mass of a teaspoonful of its matter would be one metric ton) and its surface gravity is 50,000 times greater than that of our small planet. It is probably one of the oldest stars in the Universe.

Zeta (ζ) **Piscium**, a beautiful double star, is situated almost on the ecliptic discovered by William Herschel in 1781. This system is about 140 light years away. Under good observation conditions (very clear sky and good atmospheric stability) you will be able, with a small telescope, to find a third companion, with a magnitude of 12, situated 1" from the fainter star of the pair.

Practical astronomy

Instrument of the month: the telescope

With the aid of a small telescope 2.5 inches (60 mm) in diameter, you can make excellent observations. In particular, you can repeat the discoveries of Galileo, Huygens and Cassini, who only had tools of mediocre quality available, compared with those you can buy today.

Choosing a telescope

The essential feature is the diameter: the bigger it is, the greater the amount of light it lets in, and the more possible it is, therefore, to make interesting observations. These diameters vary between 2.5 inches (60 mm) and 8 inches (200 mm). Prices of instruments increase with their diameter, but not proportionately. You can find small telescopes for about $200, including the mounting, but you need to allow about $4000 for a good 7-inch (180 mm) diameter telescope, without mounting!

Usually, telescopes are sold with several lenses, thus giving the possibility of different levels of magnification. The maximum magnification is theoretically 2.5 times the diameter of the instrument in millimeters. As a general rule, do not use over-strong magnification because you will be disappointed by the lack of clarity and luminosity. What's more, you would need to fit a mechanical drive to your telescope, in the form of a clockwork movement, to compensate for the Earth's rotation and make sure the star remains fixed in the telescope's line of sight. For example, if you choose a small-diameter telescope, do not attempt to magnify more than a hundred times. The minimum magnification is equal to the diameter in millimeters divided by 7, the figure corresponding to the diameter in millimeters of the human pupil. Below this magnification, clarity will be reduced.

Magnification depends on the focal length of the telescope and the eyepiece. You can calculate this by dividing the first by the second. Thus, a telescope with 28-inch (700 mm) focal length used with an eyepiece of 1-inch (25 mm) focal length will give a magnification of 28 times. This is already more than the magnification obtained with binoculars. Focal length is the distance separating the objective from the image it produces.

Telescope

Telescopic sight

Mounting

Tripod

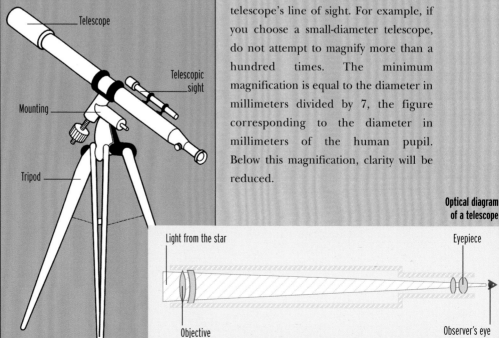

Optical diagram of a telescope

Light from the star

Eyepiece

Objective

Observer's eye

Apochromatic telescopes

A telescope's objective is usually made up of two lenses used together to reduce any chromatic aberration ▶ by limiting the dispersion of reds and blues. This makes the images much clearer than those obtained with an ordinary lens. Some very high-quality lenses are said to be apochromatic, that is to say they reduce all color dispersion (red, blue, yellow and violet) and give very good results in photographs. But they are very much more expensive: a good classic 3.3-inch (830 mm) diameter telescope costs about $1,400, but an apochromatic one of the same diameter on the same mounting would be about $2,000.

A telescope is usually fitted with an ordinary altazimuth mounting; more rarely with an equatorial mounting, which is more expensive.

▶ DEFINITION

Chromatic aberration: this is when images become iridescent due to an instrument's lens. In a simple telescope, the objective acts as a prism and disperses the colors composing light differently. You then get superimposed images of different sizes and colors. This drawback disappears in lenses with mirrors, since all the colors are deflected in the same way.

Jupiter and Saturn
The planets are seen here through a small 3-inch (80 mm) telescope.

The Pleiades
The cluster is seen here through a 2.5-inch (60 mm) telescope.

The Moon
Even a 3-inch (80 mm) telescope gives a detailed view of the Moon's relief.

What should you observe?

Mainly luminous objects in which the image quality allied to the quantity of light will give a good contrast. Small telescopes have a focal ratio (the ratio between the diameter of the objective and the focal length) that is usually higher than 12, giving them a picture quality superior to larger telescopes. Small telescopes are ideal for observing the Moon and planets and measuring double stars. They also enable you to observe the Sun, provided you take plenty of precautions and don't use the so-called "sun" filters for eyepieces that are usually sold with them. You need to use either an objective filter, which you put in front of the instrument's lens, or a projection system (see p. 216). For the beginner, these small telescopes are easier to handle than large ones, especially because the line of sight is direct: you observe what you are pointing at.

Introduction to astronomy

Crescent-shaped moon
This photo taken by the ESO shows an abundance of detail on the Moon's surface.

Discovering the Moon

The Moon is without any doubt the astronomical body that is observed most. Visible day and night, it links the stars to the Sun, and fascinates by its continually changing appearance and rapid movement across the sky. More than 2,500 years ago, the successive phases of the Moon allowed an early scientific discovery to be made. Whereas it was obvious that the Moon was round, there was no indication that it was spherical, but when people tried to reproduce its "figures" (phases), it emerged that the Moon could not be flat. The second advance was in measuring its distance from the Earth, which was done by Aristarchus of Samos in the 3rd century BC. Then, in the 17th century, Gio-Domenico Cassini finally explained why only one half of the Moon was visible: its rotation period is the same as the period it takes to orbit the

Earth, so that we always see the same side of it. Finally, in the 19th century, the "lunar theory" was established. This was a series of equations enabling all the satellite's multiple movements to be described and predicted: first, its orbit around the Earth, which is responsible for its successive phases, then its rotation on itself, which only reveals one side of it, and lastly its numerous movements, which cause the Moon to oscillate around its axis, and in turn for this to fluctuate around a median position.

The Earth-Moon system

Following Newton's research into gravity (18th century) it was supposed that the Earth was slowing down. We now know that our planet is revolving less and less quickly because its matter is not totally rigid. Moreover, the oceans and atmosphere are rubbing at its crust and thus helping to slow it down. This slowing down (a few fractions of a second per century) is leading to a distancing of the Moon from the Earth, which has been precisely measured thanks to the space missions of the last 30 years. Powerful lasers illuminate reflectors placed on the lunar soil by the American Apollo missions and Soviet probes. The reflected light returns to Earth, from where the length of the journey is timed. The distance traveled is thus calculated to within a few millimeters. These measurements have allowed the theory that the Moon is moving away from the Earth by about 3 centimeters a year to be confirmed. An unexpected consequence of this is that in just over 130 million years' time, it will no longer be possible to see a total eclipse of the Sun from the Earth, because the Moon will be too far away to mask it completely.

Lunar soil

The plateaus of craters are the oldest features of the lunar surface. They were formed when the Moon was covered with a vast ocean of magma while it was still young. The lightest elements in this material floated above the heavy elements, which sank into the depths. Cooling caused a crust to form, which was bombarded intensely with meteorites four billion years ago, and the craters date from that period. Then, for 800 million years, the

DEFINITION ▶

Escape velocity: this is the minimum velocity required to escape from another star's gravitational field.

The Moon in figures	
Average distance to the Earth	384,401 km
Distance from the perigee	356,400 km
Distance from the apogee	406,700 km
Eccentricity of orbit	0.05
Inclination of orbit over the equator	5° 8′ 43″
Sidereal period	27.32166 days
Average radius	1,738.2 km
Mass	7.35×10^{22} kg (0.012 Earth)
Density	3.34
Escape velocity ▶	2.38 km/s

The Earth in figures	
Average distance to the Sun	149,597,000 km
Distance from the perihelion	147,054,000 km
Distance from the aphelion	152,140,000 km
Eccentricity of orbit	0.02
Sidereal period	365.2564 days
Equatorial radius	6,378.164 km
Polar radius	6,356.779 km
Mass	5.976×10^{24} kg
Density	5.52
Escape velocity	11.2 km/s

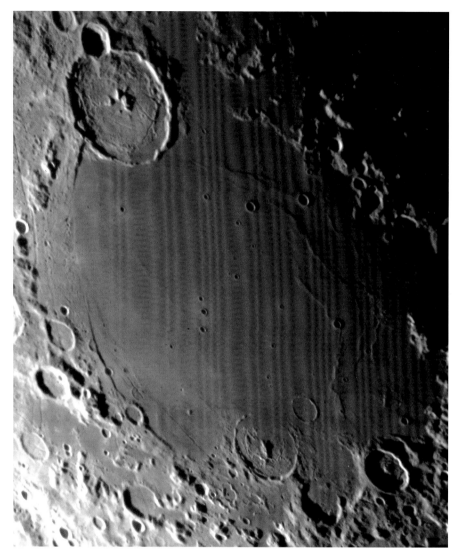

Mare Humorum (Sea of Moisture) Rising above the lunar plain, the Gassendi crater appears in fine detail.

radioactive elements present in the rocks beneath the surface heated the Moon's interior as they disintegrated. Enormous quantities of lava flowed into the great impact craters, thus forming the present "seas."

The surface is covered with a layer of dust of a thickness varying from 4 to 5 meters in the seas, and 10 meters on the plateaus. The Moon experiences seismic activity that has been recorded by measuring instruments deposited during the various lunar missions, and these are different from our earthquakes: they are less intense and probably last for longer. It is probable that these tremors on the Moon were caused by "tides" created on it by the Earth.

The lunar crust has a thickness of between approximately 50 and 100 kilometers. It is thicker on the side not facing the Earth. The Moon's core is probably composed of molten rocks and, if it has a nucleus, this must be small because the Moon does not have a large magnetic field.

The phenomenon of the tides

The Earth and the Moon, being very close together, exert an important influence on one another. The Earth has deformed the Moon through the action of "tides" and slowed it down, to the point where the Moon takes as much time to turn on itself as it does to orbit the Earth; and the Moon in turn deforms the surface of the Earth. The Earth's tides are not caused solely by the pull of the Moon; the Sun also plays a role, and in no small way, since it accounts for nearly a third of the tide-raising force.

The part of our globe that is nearest the Moon is subjected to a greater attraction than the central part, so that its bulk is pulled toward the Moon, thus resulting in a deformation. The part of the globe furthest away from the Moon is, conversely, less attracted than the central part, resulting in another deformation, which is diametrically opposed to the first. These deformations manifest themselves in the form of two large opposing waves, which go around the Earth at the same rate as the Moon, that is to say in 24 hours, 50 minutes. At any given place on Earth, therefore, the deformation occurs every 12 hours, 25 minutes. This is the tide phenomenon, experienced in all parts of the globe. It is the oceanic tides that are best known and the most spectacular, since volumes of water, being more fluid, become more deformed than solid masses.

But what one might call "tides" also exist in solid matter. The plates covering the surface of the Earth are deformed twice a day, rising and falling by about 70 centimeters with the rhythm of the Moon. The atmosphere also experiences the same phenomenon, causing variations in barometric pressure.

The presence of the Moon moving around the Earth has stabilized the axial rotation of our planet and thus spared us violent climatic variations, guaranteeing a succession of relatively regular seasons. Without this regularity and stability, life on Earth may have had more difficulty in beginning.

Precession of the equinoxes

The Earth is not a perfect sphere. Its shape is like a spheroid bulging at the equator. The pull of the Sun and Moon on such a mass causes complex movements. The attraction they exert on the equatorial bulge tends to straighten the Earth's axis perpendicularly to the plane of its orbit, just as the axis of a spinning top stands up perpendicularly to the ground on which it is turning. This phenomenon is known as precession, and it was discovered by Hipparchus in the 2nd century B.C. The direction in the sky toward which the Earth's axis points is therefore not fixed. At the present time, we are lucky enough to be able to see a star with our naked eye when we look in this direction – Polaris. But the motion of precession will turn the Earth's axis toward the star Alpha (α) Cephei in another 4,000 years, then toward Vega, in the constellation of Lyra, in 11,000 years, before coming back toward our present North Star, in Ursa Minor, in 26,000 years.

Return to the Moon?

A step on the Moon
This footprint dates from the Apollo 11 mission (1969).

Astronaut at work
Harrison Shmitt is the only scientist (a geologist) to have been to the Moon (Apollo 17 mission, 1972).

Since the space race of the 1960s, with the manned flights of the Apollo program, activity focusing on the Moon has very much slowed down. However, the automatic explorations by NASA's Clementine probe, in 1994–95, were successful in drawing up the first complete map of the Moon. Japan is also putting in place two space missions: Lunar A, planned for 2003, will analyze the soil of the Moon with excavators that will go beneath the surface to extract samples that have not been exposed to sunlight; and SELENE, planned for 2004, is to carry out an exploration in orbit and make a soft landing. The main purpose of these missions is to further our knowledge of the Moon's origins.

Meanwhile, Europe is preparing for its first lunar experiment with the SMART 1 mission that will put an experimental probe into orbit around the moon, at a date yet to be decided. It will also test out a new form of technology intended for future interplanetary voyages: Ariane 5 will put the vehicle into geostationary transfer orbit, and its height will be progressively increased by means of electrical propulsion until, after six months, it reaches the Moon. In lunar orbit, SMART 1 will carry out soil observations and experiment with a system for transmitting data to Earth. Then, at the end of its mission, electrical propulsion will be used to try to remove the probe from its lunar orbit.

Mineral wealth

There are other reasons pushing astronomers to carry out research on the Moon and to exploit it. Since it has no atmosphere, the Moon is permanently subject to solar radiation. Its soil is therefore rich in helium 3, an element deposited by the solar wind over the ages. Helium 3 is extremely rare on Earth, but nevertheless very important: it is the kind of fuel most adapted to nuclear fusion. The Moon might therefore be able to provide this fuel to satisfy the needs of our planet. Other elements could easily be extracted from the lunar soil, in particularly metals like titanium, iron, aluminum, chromium and nickel.

An astronomical site

The absence of atmosphere on the Moon makes observation considerably easier, and its surface could accommodate telescopes, enabling space to be viewed at much greater resolution. The Moon also has a unique advantage: its hidden face. Since this never sees the Earth, it is sheltered from the electromagnetic radiation produced by human activity and consequently from the electromagnetic pollution that conceals the faint signals sent out by the stars. It is the only place near to Earth that would allow very delicate observations in the radio wavelengths to be carried out. This face could thus provide a shelter for radio telescopes that might possibly receive the first signals from another civilization which had developed around a star other than the Sun.

How was the Moon formed?

This question has given rise to numerous theories. Some people have said that the Moon broke away from the Earth as it was forming; others thought that the Earth and the Moon had been formed simultaneously in the form of a double planet; and yet others speculated that it had been captured by the Earth and kept a prisoner in orbit. None of these theories fully satisfies the gravitational, mechanical and chemical data gathered from the way the Earth and Moon interact. For this reason, researchers nowadays are tending to opt for a catastrophic formation theory of the Earth-Moon system. The Earth as it was forming is said to have collided with a protoplanet with a mass equivalent to that of Mars. The collision, an extremely violent one, would have hurled out vast amounts of debris from the two bodies, which then collected in orbit around the future Earth. The disk thus created, called an accretion disk, is thought to have been responsible for forming the Moon. This theory appears to tie in with all the evidence, both mechanical and chemical. The most surprising thing about it is the speed of the process: it is all thought to have happened in only a few days! We should therefore continue to deepen our knowledge of the Earth-Moon system, both as regards its dynamics, by refining our knowledge of its movements, and as regards its chemical composition, by improving our analysis of samples taken from the Moon. All this evidence gathered together should enable us to piece together the history of the system.

Lunar relief
This picture of crater Copernicus was taken by Lunar Orbiter in 1991.

October

Lunar eclipse
During a total eclipse, the Moon's surface takes on a surprising appearance, with colors resulting from the Earth's atmospheric pollution.

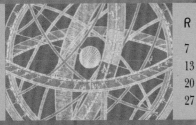

A month with your head up in the stars

Jupiter, Venus and Saturn adorn the sky during the second half of the night, and are joined by the Moon from the 17th to the 21st. The constellation of Taurus rises in the early part of the night, preceded by the splendid Pleiades cluster. The event of the month is the eclipse of the Moon, on the 27th, a sight not to be missed.

Human satellite
Astronaut Bruce McCandles left the Challenger space shuttle on February 3, 1984, to become the first "human satellite," maneuvering in space untethered to the craft.

Observations of the month

The evening sky continues to be dominated by the Summer Triangle, which is beginning its descent toward the west: along with the arrival of the Pleiades, this is a premonition of the winter nights to come. The Square of Pegasus reaches its culmination, allowing the Andromeda galaxy to be observed from the beginning of the night. To the south, Fomalhaut, in the constellation of Piscis Austrinus (popularly known as the Southern Fish), gleams on the horizon. To the north of Pegasus, the "W" of Cassiopeia, heralding Perseus and Orion, is not to be missed.

All times are given in Eastern Daylight Time (EDT) until October 31, and after that in Eastern Standard Time (EST). Calculations have been made for a northern latitude of 44° and a western longitude of 80°.

TIMES OF RISING AND SETTING FOR THE SUN, MOON AND FIVE PLANETS VISIBLE WITH THE NAKED EYE														
Date	Sun		Moon		Mercury		Venus		Mars		Jupiter		Saturn	
	Rises	Sets	Rises	Sets	Rises	Sets	Rises	Sets	Rises	Sets	Rises	Sets	Rises	Sets
10/1	7:19 a.m.	6:58 p.m.	8:05 p.m.	10:38 a.m.	6:59 a.m.	6:59 p.m.	3:43 a.m.	5:28 p.m.	6:50 a.m.	6:51 p.m.	6:39 a.m.	6:47 p.m.	1:01 a.m.	3:57 p.m.
10/5	7:24 a.m.	6:51 p.m.	10:13 p.m.	2:50 p.m.	7:21 a.m.	6:57 p.m.	3:51 a.m.	5:25 p.m.	6:47 a.m.	6:41 p.m.	6:28 a.m.	6:32 p.m.	12:46 a.m.	3:43 p.m.
10/10	7:30 a.m.	6:42 p.m.	3:20 a.m.	5:35 p.m.	7:47 a.m.	6:54 p.m.	4:02 a.m.	5:20 p.m.	6:45 a.m.	6:28 p.m.	6:14 a.m.	6:16 p.m.	12:28 a.m.	3:24 p.m.
10/15	7:37 a.m.	6:33 p.m.	9:23 a.m.	7:22 p.m.	8:32 a.m.	6:50 p.m.	4:13 a.m.	5:15 p.m.	6:42 a.m.	6:15 p.m.	6:00 a.m.	5:58 p.m.	12:10 a.m.	3:05 p.m.
10/20	7:43 a.m.	6:25 p.m.	3:14 p.m.	10:34 p.m.	8:35 a.m.	6:47 p.m.	4:23 a.m.	5:09 p.m.	6:39 a.m.	6:02 p.m.	5:46 a.m.	5:41 p.m.	11:51 p.m.	2:46 p.m.
10/25	7:49 a.m.	6:17 p.m.	5:28 p.m.	4:53 a.m.	8:59 a.m.	6:45 p.m.	4:34 a.m.	5:03 p.m.	6:37 a.m.	5:49 p.m.	5:31 a.m.	5:23 p.m.	11:25 p.m.	2:27 p.m.
10/30	7:56 a.m.	6:10 p.m.	12:28 p.m.	10:38 a.m.	9:20 a.m.	6:43 p.m.	4:46 a.m.	4:57 p.m.	6:34 a.m.	5:37 p.m.	5:17 a.m.	5:06 p.m.	11:13 p.m.	2:08 p.m.

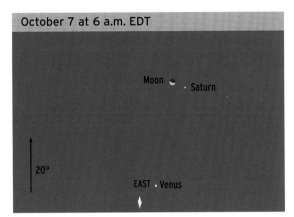

October 7 at 6 a.m. EDT

Moon • Saturn

20°

EAST • Venus

The planets

It is in the second part of the night that the planets are the most interesting. **Saturn** rises first, followed by **Venus**, then by **Mars** and **Jupiter** in the early morning. The Moon joins the parade at the end of the night from the 7th onward.

The eclipse of the Moon

The important moments of the lunar eclipse are when the Moon enters the umbra (the shadow), on the 27th at 9:14 p.m., the start and end of the total eclipse, and then its emergence from the umbra, which occurs at 11:44 p.m. While observing this event, you might take the opportunity to repeat the experiment of measuring the distance between the Earth and the Moon, as it was devised by

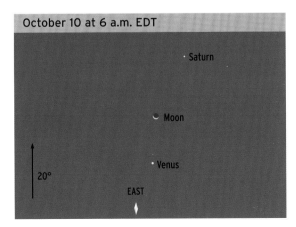

October 10 at 6 a.m. EDT

Saturn

Moon

20°

Venus

EAST

1 Friday		
2 Saturday		
3 Sunday		
4 Monday		■ 5:54 a.m. Minimum of Algol
5 Tuesday		■ 6 p.m. Moon passes the apogee: 404,328 km
6 Wednesday		■ 6:12 a.m. Last quarter of the Moon
7 Thursday		■ 2:43 a.m. Minimum of Algol ■ The Moon and Saturn move closer together. Visible at the end of the night (minimum 5.4° at 5:08 a.m.)
8 Friday		
9 Saturday		■ 11:32 p.m. Minimum of Algol ■ Maximum of the Draconids meteor shower (2 meteors an hour)
10 Sunday		■ The Moon and Venus move closer together. Visible at the end of the night (minimum 3.8° at 1:52 p.m.)
11 Monday		
12 Tuesday		■ 8:21 p.m. Minimum of Algol ■ The Moon and Jupiter move closer together. Visible at the end of the night (minimum 1.6° at 2:13 a.m.)
13 Wednesday		■ 10:48 p.m. New Moon ■ Partial eclipse of the Sun. Visible from the Northwest of America
14 Thursday		■ The Moon and Mercury move closer together. Difficult to observe in the early part of the night (minimum 0.2° at 9:16 a.m.)
15 Friday		

Aristarchus more than 2,200 years ago and improved on by Hipparchus a century and a half later. The principle of the method involves estimating the relationship between the size of the Moon and the size of the Earth's shadow. The spectacle of the lunar eclipse enables you to see the shape of the Earth's shadow, and this was one of the pieces of evidence for the Earth's spherical shape that was put forward by Aristotle in the 4th century B.C.

The eclipse of the Sun

The Sun's partial eclipse on October 13 is only visible from the extreme west of the North American continent.

Shooting stars

The **Draconids** are visible from the 7th to the 10th, with their maximum on the 9th. These shooting stars can be spectacular: in 1946, 1,000 meteors an hour were counted! The **Northern Piscids**, slow meteors that begin to light up the sky from September 25, are visible until the beginning of November. The maximum is reached around the 12th.

Finally, the **Orionids** appear from October 2 and remain visible for the whole of the month, with a maximum on the 20th. They are the remnants of the celebrated Halley's comet, and are very fast-moving meteors.

A region of the sky to explore: Taurus

Taurus, the Bull, is one of the most important zodiacal constellations and one of the most interesting to observe. It contains two very beautiful star clusters that are visible to the naked eye: the Pleiades and the Hyades. In mythology,

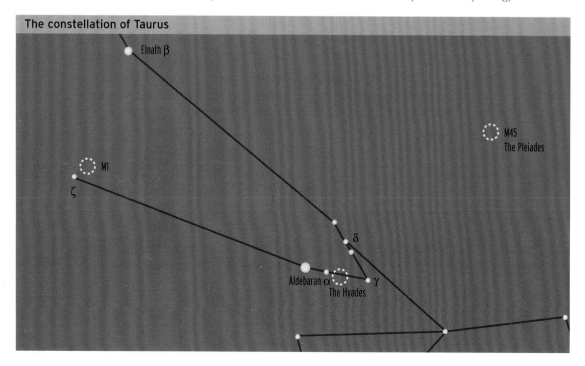

The constellation of Taurus

Elnath β

M45
The Pleiades

M1

ζ

δ

Aldebaran α
The Hyades

γ

the Bull recalls one of the metamorphoses of Zeus: being attracted by Europa, the daughter of Agenor, he approached her in the guise of a magnificent white bull. The girl climbed on his back and he carried her off into the sea and away to Crete, where they proceeded to spin out love's sweet dream at Gortyn, under a plane tree that never lost its leaves. From this union were born the three mythical kings of Crete, one of whom was Minos.

Aldebaran (or **Alpha** (α) **Tauri**) is called *al debaran* in Arabic, "the one who follows," because it appears in the sky just behind the Pleiades. This superb red giant with a magnitude of 0.68 is situated 68 light years from the Sun, which it is approaching at a speed of 54 km/s. It is a star with five solar masses, a diameter 25 times that of the Sun and a luminosity that is 200 times greater. You need to observe Aldebaran set against the other stars in the region, both with your naked eye and through binoculars, to appreciate its color, which is characteristic of red giants. It has a surface temperature of 3,720 K.

Elnath (or **Beta** (β) **Tauri**), from the Arabic *al natih*, "the butting one" (since it is located at the tip of the northern horn of the Bull), is a blue giant with five solar masses and a diameter eight times greater than the Sun's. The contrast with Aldebaran is striking, the surface temperature being 14,500 K. Elnath is 280 times more luminous than the Sun and is 130 light years away.

Gamma (γ) **Tauri** is another red giant, whose surface temperature is 4,720 K. Compare its color with that of Aldebaran by observing the field through binoculars.

16 Saturday

17 Sunday
- 8 p.m. Moon passes the perigee: 367,758 km

18 Monday

19 Tuesday
- Maximum of the Epsilon (ε) Geminids meteor shower (2 meteors an hour)

20 Wednesday
- 5:56 p.m. First quarter of the Moon
- Maximum of the Orionids meteor shower (60 meteors an hour)

21 Thursday

22 Friday

23 Saturday

24 Sunday

25 Monday

26 Tuesday

27 Wednesday
- 4:24 a.m. Minimum of Algol ■ 11:58 p.m. Full Moon
- Total eclipse of the Moon: enters penumbra 8:07 p.m./Enters umbra 9:14 p.m./Start of total eclipse 10:24 p.m./End of total eclipse 11:44 p.m.

28 Thursday
- End of eclipse of the Moon: emerges from umbra at 12:53 a.m., emerges from penumbra at 2:01 a.m.

29 Friday

30 Saturday
- Last day of Daylight Saving Time

31 Sunday

M1
The Crab Nebula, in the constellation of Taurus, is the remnant of a star explosion.

The star is 140 light years away and is 80 times more luminous than the Sun. It belongs to the Hyades cluster, just like its twin sister **Delta (δ) Tauri**, which is a bit further off (150 light years).

Zeta (ζ) Tauri is not part of the Hyades cluster; it is a distant blue giant (820 light years away), 4,400 times more luminous than the Sun. Again, it is interesting to compare its color with that of other giants in the field. The surface temperature of this star is 24,200 K.

The Crab Nebula, M1, was discovered by John Bevis in 1731, then by Charles Messier in September 1758, when he was searching for Halley's comet. M1 is the remnant of a supernova whose explosion was visible in 1054: this is borne out by observations of Chinese astronomers at the time and recorded in rock drawings discovered in Arizona. The Crab Nebula is a cloud spreading at a speed of 2,000 km/s at a distance of 3,600 light years from the Sun. At its center is a rapid-rotation pulsar, spinning 30 times per second, and it has a surface temperature reaching 100,000 K. The Crab Nebula can be seen through a small telescope or strong binoculars.

The Hyades is a magnificent cluster, and taking time to savor it through binoculars is a must. You will be able to see in it, side by side, giants of very different temperatures and become aware of its great diversity of stars. There are several hundred of them, forming a cluster that is very near to us (about 130 light years). Aldebaran does not belong

THE SKY IN OCTOBER

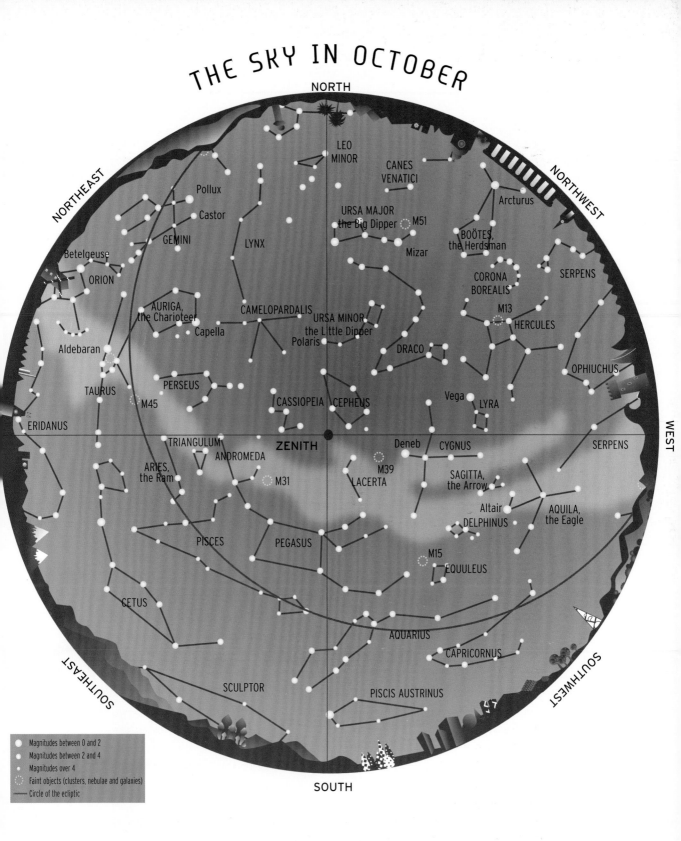

NORTH

NORTHEAST

NORTHWEST

LEO MINOR

CANES VENATICI

Pollux

Castor

Arcturus

GEMINI

URSA MAJOR
the Big Dipper

M51

BOÖTES,
the Herdsman

LYNX

Mizar

Betelgeuse

SERPENS

CORONA BOREALIS

ORION

AURIGA,
the Charioteer

CAMELOPARDALIS

M13

HERCULES

Capella

URSA MINOR
the Little Dipper

Aldebaran

Polaris

DRACO

OPHIUCHUS

TAURUS

PERSEUS

Vega

M45

CASSIOPEIA

CEPHEUS

LYRA

ERIDANUS

TRIANGULUM

ZENITH

Deneb

CYGNUS

WEST

ANDROMEDA

SERPENS

ARIES,
the Ram

M31

M39

LACERTA

SAGITTA,
the Arrow

Altair

PISCES

PEGASUS

DELPHINUS

AQUILA,
the Eagle

M15

EQUULEUS

CETUS

AQUARIUS

CAPRICORNUS

SOUTHEAST

SCULPTOR

PISCIS AUSTRINUS

SOUTHWEST

SOUTH

- Magnitudes between 0 and 2
- Magnitudes between 2 and 4
- Magnitudes over 4
- Faint objects (clusters, nebulae and galaxies)
— Circle of the ecliptic

■ How to use this chart
Hold the chart above your head, matching the word SOUTH that appears at the edge of it with the geographical south of the place you are observing from. Use a compass to help you do this.

■ This chart shows the sky that is visible at a latitude of 45°
If you are further north or further south, Polaris will be higher or lower in the sky.

■ Chart of the sky visible at 11 p.m. EDT at the beginning of the month; at 10 p.m. EDT in the middle of the month; at 9 p.m. EDT at the end of the month.

to the cluster, but is about halfway between the Hyades and us. The other bright stars of Taurus, among them Gamma (γ), Delta (δ) and Epsilon (ε), form the core of the cluster, which extends for about 8 light years.

The Pleiades center stage

The Pleiades, M45, are one of the wonders of the sky, rewarding observation with the naked eye, through binoculars and also with a telescope. The origin of the name Pleiades is uncertain. Some think that it comes from the Greek *plein*, "to sail," because, in antiquity, the cluster rose in the mornings at a time when the season was most favorable for navigation in the Mediterranean. Others think more prosaically that it comes from *pleios*, meaning "full" in the sense of numerous, on account of its appearance. The Pleiades culminate at midnight at the end of October and are associated with the ancient celebrations for the feast of the dead. Their nocturnal presence was celebrated both in Persia and in pre-Columbian America. The direction in which the Pleiades rose was also used to decide which way Greek temples should face, as well as to determine the positioning of the Pyramid of the Sun at Teotihuacan in northern Mexico.

With the naked eye, you can make out six or seven stars, but some observers claim to see up to twelve when conditions are good. The cluster, which is a few tens of millions of years old, is a rich one, having more than 500 blue stars. The most luminous of these, the blue giants, can serve as a measure when estimating the magnitude of neighboring stars. Alcyone, the most luminous in the cluster,

has a diameter 10 times greater than that of the Sun but is about 1,000 times more brilliant. But the Sun is 100 times brighter than the least luminous. The whole cluster is moving away from the solar system at about 50 km/s; the resulting apparent motion is about 5.5" per century. At this speed, it would require 30,000 years to complete an angular motion equivalent to the apparent diameter of the Moon, that is to say 0.5°.

With binoculars, the number of objects increases considerably and becomes very impressive. The nine most brilliant stars are distributed over an area of a little over 1° in diameter. It is therefore possible to observe the whole group with the aid of a small telescope fitted with a wide-angle lens. These nine stars then assume the shape of a small saucepan with a very short handle. That is why the Pleiades are sometimes confused with Ursa Minor, so it is essential to compare the two formations, and this will reveal a substantial difference in dimensions. Ursa Minor is almost 20 times larger than the Pleiades.

The Pleiades are young stars and their material is still similar to when they were formed. A great number of double and multiple stars can be seen. When you first look, Atlas and Pleione, separated by only 5', become clear through binoculars.

The Pleiades
M45 is one of the most beautiful open clusters in our sky.

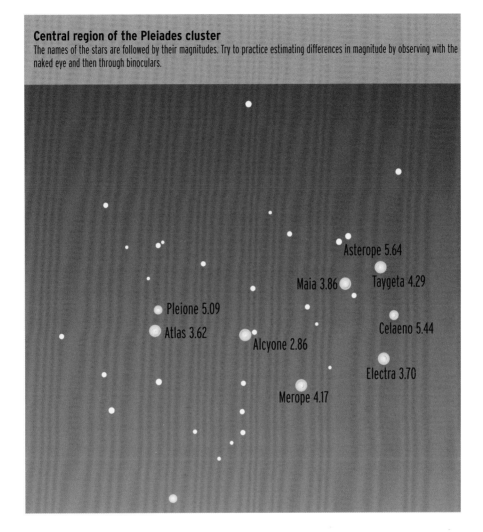

Central region of the Pleiades cluster
The names of the stars are followed by their magnitudes. Try to practice estimating differences in magnitude by observing with the naked eye and then through binoculars.

Asterope 5.64

Maia 3.86   Taygeta 4.29

Pleione 5.09

Atlas 3.62   Celaeno 5.44

Alcyone 2.86

Electra 3.70

Merope 4.17

With a small spyglass, you will discover that Atlas is a double whose components are quite difficult to distinguish (0.4" separation). The field around Alcyone is fascinating; with small optical instruments, you can make out a quadruple system. In the 1880s, progress in astronomical photography revealed nebulosities that were invisible to the naked eye and seemed to shroud the cluster. We now know that these are behind the stars in the cluster and are illuminated by the brightest among them, just as the backdrop on a stage is illuminated by the footlights. The distance from the cluster to Earth was for a long time the subject of controversy. The Pleiades are indeed too far distant for their parallax to be measurable with traditional instruments. It is now agreed that they are about 400 light years away. In other words, the twinkling that we see today comes from light that left the stars at the time Gio-Domenico Cassini was born – in 1625.

Practical astronomy

Instrument of the month: the Newtonian telescope

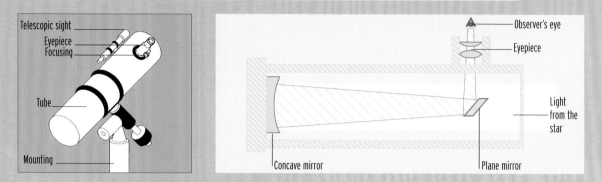

In 1671, Isaac Newton designed, and produced, a light collector that was very different from the telescopes astronomers had used since the beginning of the 17th century. The objective was made from a concave metal parabolic mirror, which concentrated the star's light toward its focus. The image was then reflected by a secondary mirror to a focal point at the side of the tube, where it was magnified by an eyepiece.

In our own day, the Newtonian telescope is one of the instruments most widely used by beginners. With a diameter usually above 4 inches (100 mm), it receives much more light than a 2.5-inch (60 mm) telescope. The most common instruments offer combinations described as 114–900, 150–900, 150–1200, etc. The first number indicates the diameter of the mirror, and the second the focal length, in millimeters. Usually, the instruments come with two or three eyepieces with different focal lengths. The magnification obtained is equal to the ratio between the focal length and the eyepiece. With a telescope of 115–900 and an eyepiece of 0.8 inch (20 mm), it reaches 900–200, that is to say 4.5. As with the smaller telescope (see

p. 162), it would be counterproductive to try to magnify too much.

How should you choose your telescope?

Once you have chosen the diameter, you should think about other aspects. While the tubes containing the optics are more or less identical, choice of accessories can make a big difference. You need to consider the following points.

The finder is the accessory that enables you to set your sight on an object in the sky and frame it in the telescope's field. There are two types that come with small models. The traditional finder is a very tiny telescope fitted with an objective and an eyepiece, generally of very average quality and quite difficult to adjust. It is preferable to opt for a finder of the "starpointer" type, without a lens, which is a simple view-finding instrument with a luminous sight attached. It is easy to regulate and can be fitted to any instrument for a price of about $50.

The eyepiece holder, as its name indicates, holds eyepieces, which are 1 or 1.25 inches (24.5 or 31.75 mm) in diameter. It is preferable to buy the larger option because the choice of focal

Optical diagram of a Newtonian telescope The observer looks at the mirror where the image is formed; the view of the sky will be reversed.

length is greater and the quality is often better, and it will be easy to transfer from one instrument to another when you decide to get a telescope with a larger diameter.

The mounting is the mechanical part that supports the telescope and enables you to direct it toward the object observed in the sky. Telescopes usually come with an equatorial mounting. If it is well positioned (see p. 163) this sort of mounting enables you to follow a star with a single movement.

Most "standard" mountings are of average quality and uncertain stability. But good quality mountings are expensive: while it is possible to find a telescope fitted out with accessories for $300, an effective mounting on its own costs at least $1,000.

In regards to the tripod, or stand, stability is the basic criterion. Avoid wooden stands as they are too flimsy. Choose an aluminum one for preference

Choice of instrument is essentially a matter of compromise related to budget. Note, though, that the same 114–900 telescope can cost from $300 to $1,200, depending on the mounting, the stand and the lens.

What should you observe?

Newtonian telescopes, whether or not they are motorized, are excellent observation instruments, as much for the Moon and planets as for faint objects - clusters, nebulae and galaxies. Their ease of use allows you to make good drawings of the surfaces of planets and of the Moon. On the other hand, it is more difficult to take photographs, except of planets (see p. 198).

First and foremost, it is essential to familiarize yourself with a Newtonian telescope. In order to capture the image, the observer does not look toward the sky but toward the mirror where the image is formed: the telescope acts like a rearview mirror! One piece of advice: to begin with, observe fixed objects on the ground, and practice pointing at them. You will get used to the reversed view resulting from the interplay of mirrors: what is on the right of the image is on the left in the sky, and vice versa. The telescopic sight, which is often quite small, can easily become disturbed. When you have mastered the telescope's movement and stabilization, you can look at the sky, beginning with the most luminous objects, like the Moon and planets, taking care not to magnify too much. It is preferable to begin with the least powerful eyepieces - 1 or 1.25 inch (25 or 30 mm) - remembering that the greater the focal length of an eyepiece, the less powerful it is! After that you can use more powerful ones, 0.35 inch (9 mm) for example, which is the ideal eyepiece for a 4-inch (100 mm) Newton.

The Moon through a telescope A 6-inch (150 mm) telescope gives a very fine view of lunar relief. Here, the Stadius crater can be seen.

Saturn through a telescope Saturn's rings are very clearly defined through a 4-inch (100 mm) telescope.

Introduction to astronomy

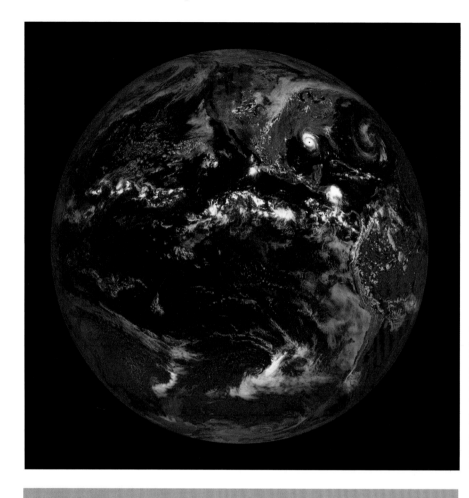

The Earth seen from space
This picture was taken by the NOAA satellite GOES 7 in August 1992. The Andrews tropical storm over Florida is clearly visible.

Discovering Planet Earth

The Earth is the third planet in order of distance from the Sun, after Mercury and Venus. It is revolving at an average distance of 150 million kilometers from the Sun, an orbit that it travels in 365 days and six hours. It is the largest of the rocky (or telluric) planets, with a diameter of 12,756 kilometers at the equator and 12,714 kilometers from north to south. This flattening out at the poles is due to centrifugal force: the Earth was revolving very fast during its early years. Today, its rotation period (length of day) has fallen to 23 hours 56 minutes.

Seen from space, the Earth is mainly white and blue: the white part represents cloud cover, whereas the blue, which occupies the major part, is due to the oceans. Its atmosphere is principally made up of nitrogen (78%) and oxygen (21%), together with argon (0.9%) and traces of carbon dioxide (0.03%), neon and other rare gases.

The concentration of water vapor varies between 1% and 3% according to region and season. Atmospheric pressure is about 1,000 hectopascals (1 bar) at sea level and diminishes rapidly with altitude: it is 100 times lower at an altitude of 30 kilometers and 1 million times lower (0.1 pascal) at an altitude of 90 kilometers. The physical conditions of temperature and pressure allow liquid water to remain at the surface of the planet, which led to the apparition of life.

The miracle planet

This liquid and gaseous environment has particular consequences for the Earth: according to our present state of knowledge, it is the only planet to harbor life. Life forms appeared in the oceans about four billion years ago as groupings of organic molecules ❱, capable of reproducing and diversifying. This did not happen by chance – liquid water is a remarkable solvent and it was this that enabled organic molecules to meet, unite and then separate to create more and more complex structures.

That life is limited to Earth might therefore be explained by the fact that our planet is the only one to have water in liquid form, a privilege due to its position in relation to the Sun, because liquid water only exists within a restricted band of temperature: between 273 K (0°C) and 373 K (100°C) (for 1 bar of pressure). Below 273 K, it is solid (ice) and beyond 373 K it is gaseous (water vapor), both states which are not favorable to the chemistry needed for life. The amount of sunshine received on Earth, at 150 million kilometers' distance from the Sun, is hardly sufficient for it to reach this ideal temperature band. In fact, without its atmosphere, the Earth would register an average temperature of 263 K (–10°C) and its oceans would be frozen. But thanks to the modest concentration of carbon dioxide in the atmosphere, it is subject to a slight greenhouse effect, which raises its temperature to 288 K (15°C).

The oceans, which regulate climate and sustain life, cover 72% of the globe's

❱ **DEFINITION**

Organic molecules: Chains of carbon atoms, which may also include atoms of hydrogen, oxygen, nitrogen, etc.

The circle of the climates

The Earth is sensitive to astronomical factors that change the amount of sunlight it receives.

First of all, in a period of 26,000 years, its axial rotation describes a cone shape in space pointing in different directions, like a spinning top gone mad. Thus, periodically, the Earth has its north pole pointing toward the Sun when it is at minimal distance from it (perihelion); the boreal summer is then particularly hot. But 13,000 years later, the north pole points toward the Sun when the Earth is furthest from it in its orbit (aphelion); at this time, the summer is less hot.

To this drift is added a variation in its angle relative to the vertical (perpendicular to the plane of the Earth's orbit). This inclination oscillates between 22° and 24° 30' according to a 40,000-year cycle. Today, it is at 23° 27'. The greater the inclination, the more the seasons are accentuated (hotter summers and colder winters).

Finally, the outline of the Earth's orbit varies between a circle and an ellipse, according to a 100,000-year cycle. When the orbit is elliptic, the amount of sunshine varies by 10% over the course of the year, between the perihelion and the aphelion.

To these various astronomical cycles are added other factors, purely terrestrial in origin (changes in ocean currents, for example), which permanently alter the climate. Human factors can be added in too: the increase of carbon dioxide in the atmosphere, caused by our industrial activities, increases the greenhouse effect and raises the temperature.

Studying volcanoes from space
This radar image was taken during an eruptive phase of the Kamchatka volcanoes in Russia (October 1994).

convection move around in its mantle, and heat is released on the surface just like water bubbling in a pot. These currents in the rock, which move at speeds of a few centimeters a year, stretch the Earth's crust when they arrive at the surface, creating fissures where magma pours out, forming the ridges of underwater volcanoes. New crust is thus created along these volcanic zones (known as "accretion") and spreads out from the fissures, as if on a conveyer belt. At the margins of the oceans, as the convection current beneath the mantle starts to dip down again, the oceanic crust above does exactly the same: the conveyor belt takes it down into the troughs (known as "subduction"), where the crust melts again into the Earth's mantle.

surface. Gravimetric and sonar studies have revealed the nature of the sea depths and enabled us to observe that the Earth's relief is dominated by oceanic basins with a median depth of −4,800 meters, whereas the median level of the continents is near to 300 meters. When it comes to the detail, this distribution is complicated by particular areas of relief: chains of underwater volcanoes at a depth of 2,500 meters; underwater trenches that plunge to −11,035 meters; and, on land, mountain chains that reach 8,800 meters in altitude.

The mountain chains are buckled, a sign of the large-scale horizontal deformation of the Earth's crust, a process that is peculiar to Earth. Similarly, the chains of underwater volcanoes, which are like a wound in the crust 60,000 kilometers long, have no other equivalent in the solar system.

Ground currents

Whereas, on other planets, evacuation of internal heat is mainly vertical (by the simple conduction of heat toward the surface), the Earth is so hot (5,300 K at the central core) that ripples of

Voyage to the Center of the Earth

The study of the seismic waves crossing the Earth tells us about its structure, divided into crust, mantle and core.

The outer crust, about 10 kilometers thick beneath the oceans and about 50 under the continents, is a "foam" of light rocks, pushed out from the underlying mantle.

The mantle is made up of denser minerals, rich in iron and magnesium. About 100 kilometers down, the increase in temperature causes this mantle to become malleable and partially to melt. Below that, the mounting pressure takes over from the heat and keeps the mantle solid, until you get down to 2,900 kilometers.

At this level, the core of molten iron begins, a metal ball that is rotating like the rest of the planet and is responsible for the magnetic field. From 5,150 kilometers to the center of the Earth (6,380 kilometers), the strong pressure compresses the core into a mass of solid iron, where the temperature is over 5,300 K.

Europe and the northern part of North Africa
The ERS1 satellite means that exceptional mapping tools can be developed.

The Earth's surface is divided by ridges, troughs and other faults into about 10 "tectonic plates," each with their own direction of movement. Situated on these plates, the continents are changing position over time; they are said to be "drifting."

The Earth's relief is permanently being modified by erosion, and in particular by the water cycle. The evaporation of seawater and its precipitation over the land mean that rocks are being attacked and are disintegrating. The products of this erosion are swept along to the sea by rivers, and deposited there in the form of sediment. This wearing away is reducing the relief by a few centimeters per thousand years, or several meters per million years. If the upthrust of the mountains were to cease, in less than a hundred million years the Earth's relief would be nothing more than a dull plain 2 meters high. However, this will not happen tomorrow: the heat stored in the terrestrial globe will maintain volcanic activity and plate tectonics for another two billion or three billion years yet.

The sky tomorrow

The Earth seen from space

Hurricane Andrew, August 25, 1992
Satellites enable meteorological phenomena to be predicted and monitored.

Satellites orbiting the Earth have revolutionized our view of the globe, and the image of a small blue planet floating in space has become familiar to us. But the satellites have above all enabled us to study the regions they fly over in a systematic way, by means of cameras and other specialized sensing equipment. The first area of research to benefit from these orbital platforms was meteorology. From 1960 onwards, the satellites Tiros and then Nimbus gathered daily pictures of cloud formations, which were soon joined by temperature measurements obtained by infrared radiometers. Later on, spectrometers were installed, to assess the distribution of radiation according to wavelength: interruption of the signal on particular frequencies ("absorption bands") meant that types of gaseous molecules could be identified (water vapor, ozone, carbon dioxide) and their concentration measured. This was how the meteorological satellites revealed holes in the ozone layers above the poles.

A new tool for geologists

Observation of the ground is just as instructive. Remote sensing satellites, like Spot and Landsat, gather pictures of the terrains they have flown over in particular wavelength bands – the blue, green, red and infrared. By combining several different versions of the same image, researchers can see the emergence of certain soil characteristics, both mineral and vegetable. Thus, geologists know that a soil rich in copper will be luminous in the green and dark in the red. And agricultural scientists are able to spot ailing plant life by the fact that their brightness is dimmer in the infrared. The satellites are in continuous orbit and fly

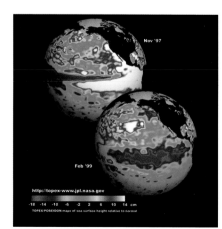

Surveillance of El Niño
The Topex/Poseidon satellite follows the evolution of temperature conditions in the oceans and contributes to the study of the El Niño phenomenon, which reaches the coast of Chile at Christmas time.

over the same region every few days, so specialists can compile thematic charts and continually update them.

The Earth is also examined by radar. Satellites like Seasat and ERS are fitted with transmitting antennae that bombard the ground with radio waves. They pick up the echoes displayed in the form of images. Received day and night, and even through cloud, these images give information about the nature of the terrain, in particular the dimensions of rough patches (sand, stones, larger rocks, etc.) and the water content. Some radar frequencies can even penetrate the ground for several meters before being reflected: thus it is possible to find out what lies beneath the ground from space.

Another application of radar waves is for altimetry. By measuring exactly the time taken by signals to travel and come back, this type of satellite can determine its distance from the ground to within a few centimeters. For example, Topex/Poseidon measures the height of oceans and reveals large water accumulations forming bulges several meters above normal sea level, and also studies the currents that flow up and down them.

The precise shape of the Earth can also be measured, because its irregularities change the local gravitational field and therefore the satellite's orbit. This is how we know that an indentation of several tens of meters affects the Earth's curvature at the level of the south pole – thus making our planet pear-shaped!

Spot satellites

Spot 5

France launched its first remote-sensing satellite, Spot 1, in 1986. The latest one, Spot 5, entered service in 2002. Spot satellites are in orbit over the poles, at an altitude of 830 kilometers. They are said to be "heliosynchronous," because they always pass over each area at the same solar time. As the light is therefore constant, the images are easier to compare. Each Spot possesses two identical cameras, pointed toward Earth: they register two separate bands of images. They can take photographs vertically (under their trajectory), or point to the side.

By programming shots, the operators can thus collect two pictures of the same region taken on two different orbits and from two different angles. By combining them, they obtain a stereoscopic image (a relief image) of the area in question. With a long-focus lens, Spot 5 can only discern details of the order of 2.5 meters on the ground, but from an altitude of 830 kilometers, it can even see cars.

November

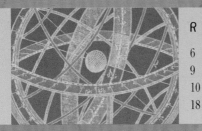

the ring planet
aturn as observed by the VLT
n December 8, 2001. It was
hen 1,209 billion kilometers
rom Earth. One of Saturn's
noons, Tethys, is visible as a
mall luminous dot underneath
ne planet.

A month with your head up in the stars

November is the month for shooting stars. This year, the Moon will be kind to astronomers, leaving the skies black at the very end of the night when the meteors will put on their display. It is also the time to observe Jupiter again, as it slowly emerges in the early morning, and especially Saturn, which this month steals the show.

Observations of the month

The Leonids
These showers of shooting stars in mid-November are one of the great attractions of the year.

At the end of the night on the 4th, the Moon passes in front of three small stars in the constellation of Cancer and finds itself about 5° away from the Beehive cluster (also known as the Manger) and 8° from Saturn. You should not miss exploring this area of the sky through binoculars – you will discover a host of stars that were first observed by Galileo with his telescope in January 1610.

The Moon will make an interesting sight as it moves through the constellation of Taurus, passing close to Elnath, the Bull's horn, on the morning of the 28th. On this occa-

All times are given in Eastern Standard Time (EST). Calculations have been made for a northern latitude of 44° and a western longitude of 80°.

TIMES OF RISING AND SETTING FOR THE SUN, MOON AND FIVE PLANETS VISIBLE WITH THE NAKED EYE														
Date	Sun		Moon		Mercury		Venus		Mars		Jupiter		Saturn	
	Rises	Sets	Rises	Sets	Rises	Sets	Rises	Sets	Rises	Sets	Rises	Sets	Rises	Sets
11/1	6:58 a.m.	5:07 p.m.	7:07 p.m.	11:41 a.m.	8:29 a.m.	5:43 p.m.	3:50 a.m.	3:55 p.m.	5:53 a.m.	4:32 p.m.	4:11 a.m.	3:59 p.m.	10:05 p.m.	1:00 p.m.
11/5	7:03 a.m.	5:02 p.m.	10:56 p.m.	2:14 p.m.	8:45 a.m.	5:44 p.m.	4:00 a.m.	3:50 p.m.	5:31 a.m.	4:22 p.m.	3:59 a.m.	3:45 p.m.	9:50 p.m.	12:44 p.m.
11/10	7:10 a.m.	4:56 p.m.	4:41 a.m.	3:57 p.m.	9:03 a.m.	5:45 p.m.	4:12 a.m.	3:44 p.m.	5:30 a.m.	4:10 p.m.	3:45 a.m.	3:28 p.m.	9:30 p.m.	12:25 p.m.
11/15	7:17 a.m.	4:51 p.m.	11:13 a.m.	7:23 p.m.	9:17 a.m.	5:49 p.m.	4:24 a.m.	3:38 p.m.	5:27 a.m.	3:58 p.m.	3:30 a.m.	3:10 p.m.	9:10 p.m.	12:05 p.m.
11/20	7:23 a.m.	4:47 p.m.	2:14 p.m.	12:31 a.m.	9:25 a.m.	5:52 p.m.	4:36 a.m.	3:32 p.m.	5:25 a.m.	3:47 p.m.	3:15 a.m.	2:52 p.m.	8:50 p.m.	11:45 a.m.
11/25	7:29 a.m.	4:44 p.m.	3:58 p.m.	6:17 a.m.	9:24 a.m.	5:52 p.m.	4:49 a.m.	3:27 p.m.	5:23 a.m.	3:35 p.m.	3:00 a.m.	2:35 p.m.	8:30 p.m.	11:25 a.m.
11/30	7:36 a.m.	4:41 p.m.	7:44 p.m.	10:25 a.m.	9:08 a.m.	5:43 p.m.	5:01 a.m.	3:23 p.m.	5:21 a.m.	3:24 p.m.	2:44 a.m.	2:17 p.m.	8:09 p.m.	11:05 a.m.

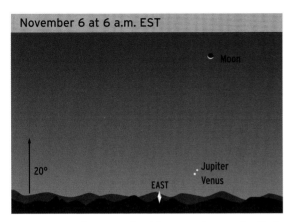

November 6 at 6 a.m. EST

Moon

Jupiter
Venus

20°

EAST

sion, you should try to observe the Crab Nebula, M1, through binoculars: at 3 a.m. it will be in the alignment of Elnath and the Moon, 5° to the south of the Moon.

The planets

The beautiful sight of **Venus** and **Jupiter** approaching one another, crowned by the Moon and **Saturn**, is offered early in the morning of the 6th. The distance separating the two planets is no more than 1° and the event takes place less than 2° away from Gamma (γ) Virginis, a beautiful star in the constellation of Virgo (magnitude 3.5). The Moon moves from Saturn to Venus between the 7th and the 10th. It can be observed quite easily, and the movements of the planets in relation to one another are clearly revealed. This is

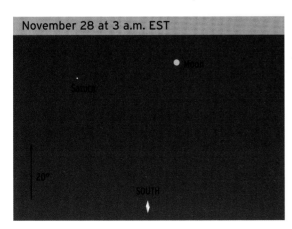

November 28 at 3 a.m. EST

Moon

Saturn

20°

SOUTH

1
Monday
- ■ 9:22 p.m. Minimum of Algol

2
Tuesday
- ■ 1 p.m. Moon passes apogee: 404,499 km

3
Wednesday
- ■ The Moon and Saturn move closer together. Visible in the middle of the night (minimum 0.5° at 3 p.m.)

4
Thursday
- ■ 5:51 p.m. Minimum of Algol

5
Friday
- ■ 12:53 a.m. Last quarter of the Moon. Venus beside Jupiter in easter predawn sky
- ■ Maximum of the Taurids meteor shower (10 meteors an hour)

6
Saturday
- ■ Venus and Jupiter move closer together. Visible at the end of the night (minimum 1° at 6 a.m.)

7
Sunday

8
Monday

9
Tuesday
- ■ The Moon and Venus move closer together. Visible at the end of the night (minimum 0.2° at 8:41 p.m.)
- ■ The Moon and Jupiter move closer together (minimum 1° at 10:40 a.m.)

10
Wednesday
- ■ The Moon and Mars move closer together. Visible at the end of the night (minimum 0.5° at 11:20 p.m.)

11
Thursday

12
Friday
- ■ 9:27 a.m. New Moon
- ■ Maximum of the Alpha (α) Pegasids meteor shower (3 meteors an hour)

13
Saturday
- ■ The Moon and Mercury move closer together. Visible in the early part of the night (minimum 1° at 2:19 a.m.)

14
Sunday
- ■ 9 a.m. Moon passes the perigee: 362,313 km
- ■ Maximum of the Andromedids meteor shower (10 meteors an hour)

15
Monday

the moment to have a go at taking photographs without a telescope. With a simple camera, try to record from day to day the area of sky where the planets are. Exposures of about 10 seconds, with a focal length of 50 mm, will be enough to discover what wonders await you.

Saturn takes center stage

Saturn rises a few hours after sunset and remains visible all night. This is also the time when the Cassini spacecraft will release the Huygens probe so that it can land on Titan next January 14. Observation conditions are particularly good; optimum conditions will be reached on January 13, 2005, but then the inclination of the ring will be slightly less favorable than at the opposition of 2003. However, to observe the planet, you need to have a small instrument available, or strong binoculars mounted on a tripod. It is only with this type of instrument that you will be able to see the ring, together with Titan, the largest of its satellites. When the first observations were made, in 1610,

The rings of Saturn
These are made up of a host of small particles.

Night on Saturn
This picture was taken when the Voyager 1 probe was flying past the planet.

Galileo was unable to see the ring. As he only had a poor-quality telescope, he thought that the planet was triple, accompanied by two very close companions. Forty years later, Christian Huygens, equipped with a much better-quality telescope, observed that the ring was not touching the planet and that Saturn possessed a moon, Titan. You can follow Titan's movements from day to day: it revolves around Saturn in a little under 16 days.

Conditions are very favorable for admiring Saturn's rings: their inclination of 27° relative to the ecliptic is almost at its maximum in relation to the Earth. They are therefore very open. With a 2.5-inch (60 mm) telescope, you can already see Cassini's Division, that black band separating the ring into two distinct parts.

Saturn in figures	
Distance to the Sun	9.539 AU, 1,433,530,000 km
Distance from the perihelion	9.01 AU, 1,352,550,000 km
Distance from the aphelion	10.004 AU, 1,514,500,000 km
Eccentricity of orbit	0.06
Inclination of orbit	2.49°
Equatorial diameter	120,536 km
Mass	5.69 10^{26} kg (95 times the Earth)
Revolution period	29.46 years
Orbital velocity	9.09 km/s
Rotation period	10.39 hrs
Escape velocity	35.6 km/s

16 Tuesday

17 Wednesday
■ Maximum of the Leonids meteor shower (1,000 meteors an hour)

18 Thursday

19 Friday
■ 12:59 a.m. First quarter of the Moon
■ 1:55 a.m. Minimum of Algol

20 Saturday
■ Midnight Greatest eastern elongation of Mercury (22.2°)

21 Sunday
■ 10:44 p.m. Minimum of Algol

22 Monday

23 Tuesday

24 Wednesday
■ 8:33 p.m. Minimum of Algol

25 Thursday

26 Friday
■ 3:08 p.m. Full Moon

27 Saturday

28 Sunday

29 Monday

30 Tuesday
■ 6 a.m. Moon passes apogee: 405,953 km
■ The Moon and Saturn move closer together. Visible at the end of the night (minimum 5.5° at 9:25 p.m.)

Inclination of the rings
Between 1996 and 2000, the Hubble Space Telescope was able to measure the evolution of the inclination of Saturn's rings.

Ring A, the outer one, is 20" from the planet, the inner ring is 10.7" away, and the center of the Cassini Division is 17.2" away. Toward the outside of the ring, 19" rom the sphere, is the Encke Division, much less obvious than the Cassini Division: in order to see it, you need telescopes of at least 6 inches (150 mm) in diameter, and good conditions. The ring's detail is much easier to discern with yellow or orange filters, which show up the differences in color and contrast marking its various zones.

Shooting stars

Several meteor showers arrive one after the other in November: the **Taurids**, the **Alpha** (α) **Pegasids**, the **Andromedids** and above all the **Leonids**, which are generally most spectacular with a very pronounced maximum on the 17th at the very end of the night.

They are caused by the comet Temple–Tuttle, whose 33-year period brought it back into the vicinity of the Sun at the end of February 1998. During the maximum, on the 18th, which often lasts for only a few minutes, it is possible to see more than one meteor per second. It is an impressive sight, but you have to be patient and also face the rigors of the November nights, which are often cold. The Leonids maximum was particularly spectacular in 1933, then again in 1996. In 2002, the maximum was very pronounced with several meteors per second for a few minutes around 5 a.m.

THE SKY IN NOVEMBER

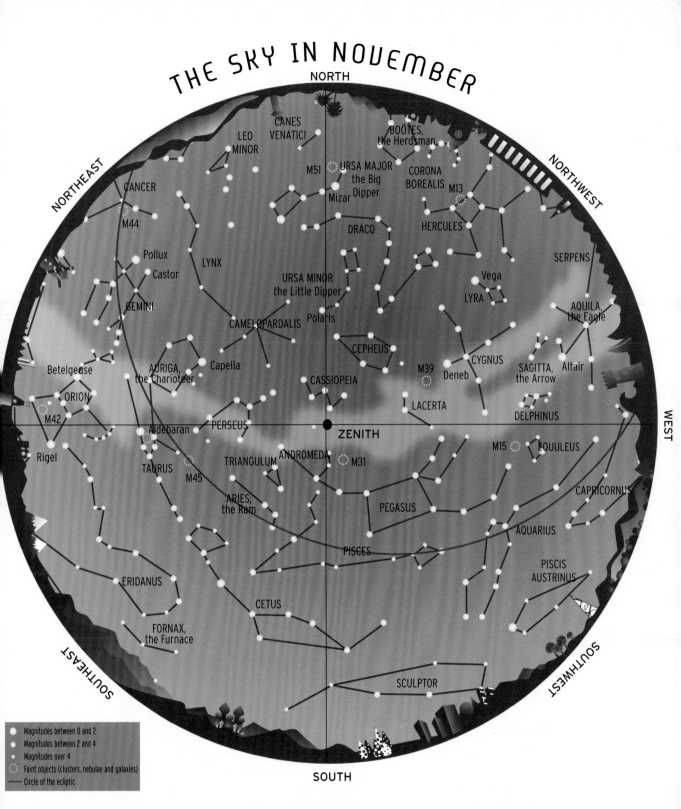

NORTH

NORTHEAST

NORTHWEST

CANES VENATICI

LEO MINOR

BOÖTES, the Herdsman

M51

URSA MAJOR
the Big Dipper

CORONA BOREALIS

M13

CANCER

Mizar

M44

HERCULES

DRACO

Pollux

LYNX

SERPENS

Castor

URSA MINOR
the Little Dipper

Vega

GEMINI

AQUILA,
the Eagle

CAMELOPARDALIS

LYRA

Polaris

Betelgeuse

CEPHEUS

CYGNUS

SAGITTA,
the Arrow

Altair

AURIGA,
the Charioteer

Capella

M39

Deneb

ORION

CASSIOPEIA

LACERTA

DELPHINUS

M42

ZENITH

WEST

Rigel

Aldebaran

PERSEUS

M15

EQUULEUS

ANDROMEDA

M31

TRIANGULUM

TAURUS

M45

CAPRICORNUS

ARIES,
the Ram

PEGASUS

AQUARIUS

PISCES

PISCIS
AUSTRINUS

ERIDANUS

CETUS

FORNAX,
the Furnace

SOUTHEAST

SCULPTOR

SOUTHWEST

SOUTH

- ○ Magnitudes between 0 and 2
- ● Magnitudes between 2 and 4
- • Magnitudes over 4
- ◌ Faint objects (clusters, nebulae and galaxies)
- — Circle of the ecliptic

■ How to use this chart

Hold the chart above your head, matching the word SOUTH that appears at the edge of it with the geographical south of the place you are observing from. Use a compass to help you do this.

■ This chart shows the sky that is visible at a latitude of 45°

If you are further north or further south, Polaris will be higher or lower in the sky.

■ Chart of the sky visible at 11 p.m. EST

at the beginning of the month; at 10 p.m. EST in the middle of the month; at 9 p.m. EST at the end of the month.

A region of the sky to explore: Auriga, the Charioteer

The constellation of Auriga shares a star with Taurus: this is Elnath, which forms the tip of the Bull's left horn and the right foot of the Charioteer. The constellation shows the charioteer holding the goat, Amalthea, who acted as nurse to Zeus after his mother managed to get him away from his monster of a father. This is why Auriga's main star is called Capella, "the she-goat," in Latin. The charioteers of antiquity were actually drivers of coaches, vehicles that transported goods and people, but they were also responsible for keeping guard over the breeding animals that they moved around during the course of their journeys.

Capella (or **Alpha** (α) **Aurigae**), the "she-goat," is sixth in order of the most brilliant stars, and the most northerly of these. Its magnitude is 0.1. Situated 45 light years away, it radiates a luminosity 160 times stronger than that of our Sun.

Beta (β) **Aurigae** is also called Menkalinan, from the Arabic *al mankib dhi'i'inan*, "the shoulder of the one holding the reins." It is an eclipsing variable whose variability is difficult to detect for a layman, but it is still an excellent one to practice on if you want to test your ability to estimate magnitudes. Its period is 3.9 days.

Epsilon (ε) **Aurigae** is one of the most famous stars in the sky. This is another eclipsing variable of a particular type. Every 27 years, its principal component is masked by an obscure companion that

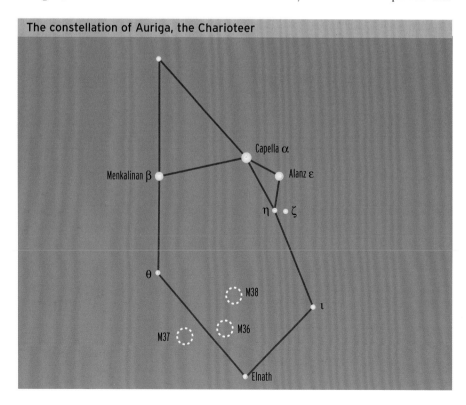

The constellation of Auriga, the Charioteer

Capella α

Menkalinan β

Alanz ε

η   ζ

θ

M38

ι

M37   M36

Elnath

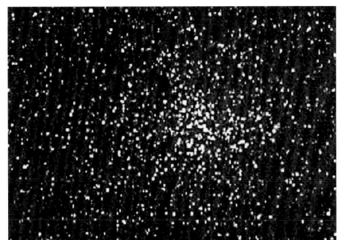

M36
One of the galactic clusters in the constellation of Auriga.

could be an immense cloud of pre-stellar matter, in which stars are still to be born. This troublemaker creates an eclipse that lasts one whole year, causing its magnitude to go from 3 to 3.8; the last one occurred in 1982–84, and the next is predicted for 2009. This main star, which produces all the light for the pair, is 60,000 times more luminous and 180 times bigger than the Sun, shining 3,000 light years away.

Zeta (ζ) Aurigae, Sadatoni or Saclateni, from *al said al thani,* "the second arm," is another eclipsing variable with a period of almost three years.

Clusters in the Charioteer

The constellation of Auriga contains three objects from Messier's catalog that you can practice locating. These are M36, M37 and M38, three very different open clusters.

M36 is very young and contains blue stars comparable to the Pleiades. However, it is 10 times further away than the Pleiades (4,000 light years)

M37 is even further away than M36. This splendid object contains more than

150 stars with a luminosity of over 12.5 in magnitude. Its diameter is about 25 light years.

M38, whose position is 4,300 light years away, is as luminous as 900 Suns.

Take your binoculars and practice observing the three clusters together, because it is an impressive sight: they are almost aligned over about 5°, that is to say 10 times the diameter of the Moon. They appear in the order M37, M36 and M38, about halfway along the line from Theta (τ) Aurigae to Elnath.

The region containing Auriga, Taurus and Perseus is one of the richest in the sky. It brings together a whole range of celestial objects as diverse as open clusters and spectacular variable stars. This beautiful sight is enhanced by the presence of Saturn, which is nearby, in Taurus, during the whole of this month. Look at it before the almost full Moon enters the region around the 21st, and dims all its rivals by its brilliance.

How to take photographs with a telescope or spyglass

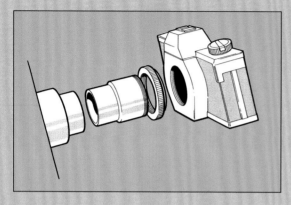

An adapter
The camera body is linked to the telescope or spyglass with a T2 ring.

Photos of the sky taken with large telescopes or from space probes have familiarized us with stunning pictures of it. Without attaining this degree of perfection, you can nevertheless take very good photographs with the aid of amateur equipment, a small telescope or spyglass.

There are two types of photographs used in astronomy. Pictures of planets belong to traditional photography: the image obtained corresponds to what you see in the instrument. Photographing what is invisible involves subjecting an area of the sky to a long exposure in order to reveal objects that are not observable through the telescope. In both cases, and however long the exposure time, your job will be made easier if the telescope is fitted with a mechanical drive.

Taking photographs through an instrument essentially means using the lens as a telephoto lens. There are three methods of doing this.

Direct photography only uses the lens of the astronomical instrument to obtain pictures. The camera body is placed directly at the point of focus, in place of the eyepiece.

Projection photography is like visual observation: the focal image is magnified by an eyepiece and projected onto photographic paper or onto a screen that can be photographed.

Finally, afocal photography means using the eyepiece and putting the camera body, fitted with its own lens, in place of the observer's eye: set to infinity, it records the image that you see through the eyepiece.

In all cases, you should adapt the photographic equipment to the telescope with the aid of a T2 ring, which costs around $25; and don't forget to restabilize the instrument once it is fitted with the photographic equipment.

Photographs of the Sun, Moon and planets

For the Sun, it is essential to use a filter over the objective, that is to say one which is placed at the telescope's aperture, not an eyepiece filter which runs the risk of breaking during the photo shoot, due to heating.

The Moon
This photograph was taken with an 3-inch (80 mm) spyglass.

For photographing planets, a colored filter, either blue or green, increases the image contrast in bands of cloud and structures visible on the surfaces of Mars and Jupiter.

Pictures of planets require a very fine-grained emulsion with a high contrast. The Technical Pan or Tmax black and white emulsions are perfectly suited to this. Each film costs about $5 to $10. If you master black and white photographs effectively, you will then be able to move on to color.

Photographing faint objects

To take pictures of galaxies, nebulae and clusters, use a combination of a wide aperture and a long exposure time. Telescopes of the Schmidt or Schmidt/Cassegrain variety are more suitable than Newtonian telescopes or spyglasses. The quality of the guides, and therefore the setting into position, is of prime importance (see p. 54).

The length of exposure necessary for photographing faint objects brings certain limitations. Because the sensitivity of photographic emulsion decreases with exposure time, you need to use emulsions that are specially designed for long exposures; however, they are extremely expensive, about three times more than ordinary emulsions, and they are also delicate to handle.

Digital imagery

Electronic imagery is developing in astronomy and tending to replace traditional film photography. The electronic version has several major advantages: the result is almost instantaneous and it is possible to repeat observations until you get total satisfaction. In addition, receptors are becoming more and more sensitive, eliminating the problems linked to long exposures. There is also a distinction between direct or projected imagery (using an electronic sensor placed at the point of the focus of the instrument) and afocal systems (placing a digital camera where the eye would be). The first method requires special sensors, CCD (Charge Coupled Device) cameras, which demand much observation experience and good quality equipment. A complete CCD camera costs at least $1,300, and the top-of-the-range for amateurs goes up to $13,000. You need to add to this the computer that enables you to process and look at the pictures, as well as the filter set enabling you to interpret the colors.

Afocal imagery is much less expensive. For about $100, you can find adapters which allow you to place a simple family digital camera behind the eyepiece of a telescope or spyglass, and take live astronomical pictures which you then just need to process and view on a computer. This is certainly the best way to approach astronomical digital imagery.

Digital image of the America Nebula

M31, the Andromeda galaxy
This image was obtained through a 6-inch (150 mm) telescope.

Discovering meteorites and shooting stars

Shooting stars have fascinated people since antiquity. Some saw divine portents in their fleeting appearances. It was necessary to wait until the 19th century to begin to understand the origin of these meteors. Shooting stars are not in fact stars; their ephemeral character and apparent velocity of motion enabled observers to comprehend that they were very nearby. We now know they are particles of dust that are consumed as they enter the Earth's atmosphere.

Dust particles from comets

These dust particles originate from comets. Comets are in fact large balls of frozen dust measuring from a few kilometers to a few tens of kilometers in diameter. Most are moving over extremely long trajectories and never go near the Sun. However, under the effect of gravitational disturbances, it occasionally happens that one of them is hurled toward the central region of the solar system. Its temperature then rises as it approaches the Sun. The ball

A bolide
This meteor, which appeared at the time of the Leonids, was photographed in November 2000.

of frozen dust starts partially to melt, releasing an abundance of water as well as the particles included in the material of the comet's nucleus. The dust particles thus released remain on the comet's flight path. The Earth, as it moves around the Sun, crosses the flight paths of comets each year and thus comes into contact with these dusty regions of interplanetary space. Each of these encounters is accompanied by the appearance of shooting stars, caused by the impact of dust particles on the upper atmosphere. These stars seem to be falling toward us, but in reality it is the Earth's atmosphere that hits these dust particles floating around in space.

From meteor to bolide

The dust particles responsible for shooting stars measure a few millimeters, even less, in diameter. The smallest ones do not leave any trace visible from Earth; those that are less than 0.1 millimeter remain intact when they encounter the atmosphere and descend slowly toward the surface of our planet. Only the trail of hot gas is visible from the ground. These luminous phenomena, known as "meteors," are caused by particles called "meteorites" (see inset paragraph p. 203). When the speck of dust is bigger, in other words, weighing a few grams or even a few kilograms, the heating of the atmosphere causes a bolide to appear. This is a luminous ball of variable color, often green but sometimes yellow, which illuminates the sky and can be as bright as full daylight just for a few seconds.

It is even possible for bolides to be visible during the day. These appearances are more rare than those of

Table of the main meteor showers			
Name	**Period of visibility**	**Meteors/h**	**Parent comet or asteroid**
Quadrantids	January 1-5	120	96P/Machholz 1 and 14911
Lyrids	April 16-25	20	C/Thatcher (1861 G1)
Eta (η) Aquarids	April 19-May 28	50	IP/Halley
Arietids	May 29-June 19	Diurnal	96P/Machholz 1 and 14911
Zeta (ζ) Perseids	June 1-17	Diurnal	2P/Encke
Beta (β) Taurids	June 7-July 7	Diurnal	
Alpha (α) Capricornids	July 3-August 19	150	
Delta (δ) Aquarids	July 15-August 28	20-25	96P/Machholz 1 and 14911
Perseids	July 17-August 24	100	109P/Swift-Tuttle
Kappa (κ) Cygnids	August 3-31	5	
Taurids	September 15-November 25	10	2P/Encke
Orionids	October 2-November 7	25	1P/Halley
Draconids	October 6-10	200	21P/Giacobini-Zinner
Leonids	November 14-21	25	55P/Tempel-Tuttle
Geminids	December 7-17	110	(3200) Phaeton
Ursids	December 17-26	20	8P/Tuttle
Each month, our section "The sky from day to day" tells you the exact moment when the maximum shower activity is forecast, if they are observable			

Comet Hyakutake
It was possible to see this comet in 1996.

the usual shooting stars, and are often accompanied by a fall of meteorites onto the Earth's soil.

Original material

The dust particles that give birth to shooting stars are of great interest to astronomers. It is in fact thought that comets might be contemporary to the formation of the solar system, 4.7 billion years ago. Consequently, analyzing these dust particles is in a way like studying the material from which the Sun and its neighbors appeared. But even though some fragments reach the ground in the form of meteorites, the smallest of them are extremely difficult to find among all the particles cluttering our environment. However, samples of meteoric material can be obtained when the Earth is crossing a comet's trajectory. Small paraffin screens, spread out over the wings of planes flying at very high altitude, gather these very fine particles, which are then analyzed on Earth. Their chemical composition resembles that of meteorites, the most volatile components having disappeared in the heating caused by their arrival in the Earth's atmosphere. Thus there are silicate compounds, metals like nickel and iron, and carbon compounds.

Meteorites

A meteorite is a natural object of extraterrestrial origin that has hit the Earth's surface. There are three main categories of these: siderites essentially contain metal (iron and nickel); stony meteorites mainly contain minerals (silicates); siderolites (stony-iron meteorites) are made up of a combination of metals and minerals. The latter are divided into chondrites if they contain chondrules (mainly silicates), and achondrites if they do not contain them. Finally, depending on the physical and chemical changes they have undergone upon arrival in the atmosphere, there is a distinction between differentiated meteorites (melted) and undifferentiated ones (those that have remained in their original state).

Out of the 200,000 metric tons or so of meteoric material attracted by the Earth, 1% reaches the ground in the form of meteorites, 20% remains suspended in the atmosphere in the form of very fine dust particles, and the rest is sent back into interplanetary space. Before entering the atmosphere, the meteoroid, which will fragment into meteorites, has an average mass of 100 kilograms. The heaviest meteorite (60 metric tons) was found in Hoba, in South Africa, in 1920. The average mass recorded for siderites is 15 kilograms, and that of stony meteorites is 3 kilograms.

Among the 22,000 meteorites cataloged, the 19 meteorites originating from Mars or the Moon were torn away from their planet because of impacts caused by asteroids or comets. The others are of interplanetary origin.

A meteorite
This 200 gram siderite was picked up in the Sahara in December 2001.

Detail of a meteorite
This speck of meteor dust was gathered at very high altitude by NASA's U2 plane. Its size is less than a tenth of a millimeter.

Could a bolide crash into the Earth?

While the Earth often encounters comet dust, it also happens that it crashes into much larger objects, like asteroids or comets themselves. It still bears the scars of these impacts, which are fortunately rare, in the form of craters several kilometers in diameter, even if erosion has made their marks less obvious. The violence of these collisions can be seen on the Moon's surface where, in the absence of atmosphere, the craters have remained intact. Collisions are common in the solar system and help to mold it through an exchange of materials between its components. Thousands of objects bump into one another around the Sun, creating smaller stars and modifying the relief of the larger planets.

Live coverage of a catastrophe

Between July 16 and 22, 1994, 21 observable fragments, some of them measuring up to 2 kilometers, crashed into Jupiter.

It was the comet Shoemaker–Levy 9, which had broken up on its previous voyage near to Jupiter. For the first time, such an accident, which seems to be relatively frequent on Jupiter (about once a century), had been predicted by astronomers and could be observed live.

The hunt for comets and asteroids

For a few years now, research into asteroids and comets has grown to enable us to get to know them better and detect those that might cross the Earth's orbit. Thus, the first asteroid was discovered in 1801, the 1,000th was listed in the catalog in 1923, the 5,000th in 1987 and the 10,000th in 1997. Today, we can number more than 60,000 of them, because the majority of moving objects are now discovered automatically. In fact, electronic records enable us to compare hundreds of images in real time and to distinguish those that are moving in relation to the stars.

The main impact craters on Earth			
Name	Location	Diameter (in km)	Age (years)
Chicxulub	Mexico	260	65 million
Vredefort	South Africa	140	1,970 million
Manicouagan	Canada	100	214 million
Acraman	Australia	90	590 million
Kara-Kul	Tajikistan	52	< 10 million
Charlevoix	Canada	46	360 million
Araguainha Dome	Brazil	40	< 250 million
Clearwater Lake West	Canada	32	290 million
Azura	Spain	30	< 130 million
Ries	Germany	24	15 million
Rochechouart	France	23	160 million
Meteor Crater	United States	1.2	50,000

The most effective instrument is a small 1-meter telescope set up in the New Mexico desert, which located more than 158,000 new asteroids and 82 comets between 1997 and 2002. This LINEAR experiment (Lincoln Near Earth Asteroid Research) is being conducted by the Lincoln laboratory at the Massachusetts Institute of Technology.

Earth Watch

Recent research puts the number of asteroids of more than 1 kilometer across likely to cause serious damage in the event of an impact with the Earth at 700,000. Among these, a few hundred are classified "at risk" because of their trajectory. The risk of collision is at present estimated at one impact every 5,000 years. Impacts of larger objects, of the order of 10 kilometers in diameter, have a probability of one every 100 million years.

Asteroids

Eros
The NEAR probe has been orbiting the asteroid Eros for a year. Despite its small size (33 km), Eros is not protected from collisions.

Asteroids are classified into several large families. The belt asteroids move between Mars and Jupiter, and are the most numerous. Some of these approach the Earth's orbit, and they are therefore called Near-Earth Asteroids (NEAs). Estimates vary, but there are thought to be about 200,000 NEAs of more than 100 meters in diameter; because of their size they resist entering the Earth's atmosphere. Between 1,000 and 2,000 of them are believed to be bigger than 1 kilometer. The other families of asteroids are further away from our planet. The Trojans, who bear the names of heroes of the Trojan war, are gathered into two groups in the orbit of Jupiter, at 60° to either side of the planet. The Centaurs, which are very distant, are revolving further away than Saturn, and even than Uranus and Neptune; they are probably related more to the comets of the Kuiper Belt than to the minor planets of the asteroid belt.

Impact on Jupiter
The comet Shoemaker–Levy 9 crashed into Jupiter in July 1994.

An impact crater on Earth
The Aorounga crater (13 km in diameter) in Chad is the remains of an impact that occurred 350 million years ago.

December

Seen from the sky
The eruption of Etna, November 3, 2002. This picture was obtained by ASTER (Advanced Spaceborne Thermal Emission and Reflection Radiometer). The plume of ashes is drifting south-southwest, over the town of Catania.

RENDEZVOUS IN THE SKY

A month with your head up in the stars

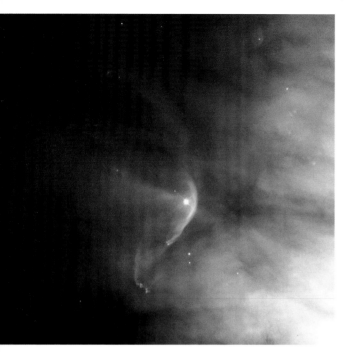

December is the month of the winter solstice, and the best observation times for astronomers are now beginning, with nights that last for more than 16 hours and a fascinating sky, featuring the constellations of Orion and Taurus. The Pleiades, Perseus, Auriga and Cassiopeia embellish the winter nights, punctuated by the appearance of the twinkling star, Sirius.

Nursery of stars
Play of light and matter at the center of the Orion Nebula. It is in these filaments and concentrations of dust that stars are born.

Observations of the month

Look hard toward the sunrise this December 23, because it is the most southerly direction possible. Compare it with the direction of the sunrise on the day of the summer solstice, June 21: you have before you a range of directions in which the Sun rises, which just goes to show how mistaken it is to say that the Sun rises in the east! The only two days in the year when the Sun rises in the east and sets in the west are the days of the equinox.

Algol has its minimum at the start of the night of the winter solstice.

All times are given in Eastern Standard Time (EST). Calculations have been made for a northern latitude of 44° and a western longitude of 80°.

TIMES OF RISING AND SETTING FOR THE SUN, MOON AND FIVE PLANETS VISIBLE WITH THE NAKED EYE

Date	Sun		Moon		Mercury		Venus		Mars		Jupiter		Saturn	
	Rises	Sets	Rises	Sets	Rises	Sets	Rises	Sets	Rises	Sets	Rises	Sets	Rises	Sets
12/1	7:37 a.m.	4:41 p.m.	7:40 p.m.	11:46 a.m.	9:02 a.m.	5:40 p.m.	5:04 a.m.	3:22 p.m.	5:21 a.m.	3:22 p.m.	2:41 a.m.	2:13 p.m.	8:05 p.m.	11:01 a.m.
12/5	7:41 a.m.	4:40 p.m.	12:00 a.m.	1:20 p.m.	8:31 a.m.	6:18 p.m.	5:14 a.m.	3:20 p.m.	5:19 a.m.	3:14 p.m.	2:29 a.m.	1:59 p.m.	7:48 p.m.	10:44 a.m.
12/10	7:46 a.m.	4:39 p.m.	6:08 a.m.	3:19 p.m.	7:36 a.m.	5:42 p.m.	5:27 a.m.	3:17 p.m.	5:17 a.m.	3:03 p.m.	2:13 a.m.	1:41 p.m.	7:27 p.m.	10:24 a.m.
12/15	7:50 a.m.	4:40 p.m.	11:27 a.m.	7:35 p.m.	6:44 a.m.	5:06 p.m.	5:40 a.m.	3:15 p.m.	5:15 a.m.	2:53 p.m.	1:57 a.m.	1:23 p.m.	7:02 p.m.	10:03 a.m.
12/20	7:53 a.m.	4:42 p.m.	1:19 p.m.	1:53 a.m.	6:14 a.m.	3:43 p.m.	5:52 a.m.	3:15 p.m.	5:14 a.m.	2:43 p.m.	1:40 a.m.	1:05 p.m.	6:41 p.m.	9:43 a.m.
12/25	7:55 a.m.	4:45 p.m.	3:45 p.m.	7:24 a.m.	6:05 a.m.	3:28 p.m.	6:04 a.m.	3:16 p.m.	5:12 a.m.	2:34 p.m.	1:24 a.m.	12:46 p.m.	6:19 p.m.	9:22 a.m.
12/30	7:56 a.m.	4:49 p.m.	7:32 p.m.	10:43 a.m.	6:09 a.m.	3:22 p.m.	6:16 a.m.	3:20 p.m.	5:10 a.m.	2:25 p.m.	1:07 a.m.	12:28 p.m.	5:53 p.m.	8:57 a.m.

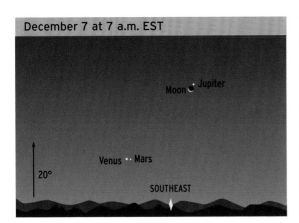

December 7 at 7 a.m. EST

Moon · Jupiter

Venus ·· Mars

20°

SOUTHEAST

The planets

In addition to Saturn, which is on show at this time of year, it is becoming easier to observe **Jupiter**, although you still need to wait until the second half of the night. Jupiter and the Moon come very close to one another on the 7th at 5:40 a.m. This encounter takes place in an area rich in stars, less than 5° from Gamma (γ) Virginis, in the constellation of Virgo. Arm yourself with binoculars to take full advantage of the spectacle: an hour later, **Mars** and **Venus** join the dance, just above the southeast horizon. On the 8th, early in the morning, you can see a very beautiful alignment of Jupiter, the Moon, Mars and Venus. It is now Spica, the main star in the constellation of Virgo, which is in the immediate proximity of the Moon.

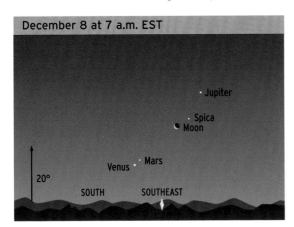

December 8 at 7 a.m. EST

· Jupiter

· Spica
· Moon

Venus · Mars

20°

SOUTH   SOUTHEAST

1 Wednesday

2 Thursday

3 Friday

4 Saturday
■ 7:53 p.m. Last quarter of the Moon

5 Sunday

6 Monday

7 Tuesday
■ The Moon and Jupiter move closer together. Visible at the end of the night (minimum 0.4° at 5:40 a.m.)

8 Wednesday
■ Alignment of the Moon, Jupiter, Mars and Venus. Visible at the end of the night

9 Thursday
■ The Moon and Mars move closer together. Visible at the end of the night (minimum 2.1° at 7:22 p.m.)

10 Friday
■ The Moon and Venus move closer together. Difficult to observe at the end of the night (minimum 3.5° at 12:09 a.m.)

11 Saturday
■ 10:30 p.m. New Moon

12 Sunday
■ 12:27 a.m. Minimum of Algol
■ 5 p.m. Moon passes the perigee: 357,986 km

13 Monday
■ Maximum of the Geminids meteor shower (60 meteors an hour)

14 Tuesday
■ 9:17 p.m. Minimum of Algol

15 Wednesday

A month with your head up in the stars

Sirius takes center stage

The month of December is an ideal time to admire Sirius, the brightest star in the sky and one of the nearest to the Sun (less than 9 light years). It is so bright that it is possible to see it change color when it rises and sets. The phenomenon is visible with the naked eye, but is more spectacular through binoculars, and it is caused by chromatic scintillation ◗.

Sirius probably owes its name to this phenomenon of scintillation, a name which is derived from the ancient Greek *seiros*, "twinkling." It was also known as the Dog Star, the dog in question being one of the two who follow Orion the hunter – the second one is Procyon – in the constellation Canis Minor.

Sirius is a medium-sized star (2.35 times the Sun's mass) and hot, with a surface temperature of the order on 10,000 K. From the beginning of the 19th century, its movements were carefully studied, and deviation from its trajectory seemed to indicate the presence of an invisible companion. This was found in 1862, during observations using the largest lens in existence at the time, and it turned out to be a very small, not particularly bright, star, orbiting Sirius in a period of about 50 years and at a considerable distance from it – 3 billion kilometers, to be exact, which is as much as the distance separating Uranus from the Sun. The density of this object is 40,000, and it is a white dwarf, the skeleton of a star that has completed its evolution.

▶ DEFINITION

Chromatic scintillation: the Earth's atmosphere acts as a prism and disperses light from the stars. When the star is very bright, like Sirius, there is enough energy in the dispersed colors for these to be visible. The turbulence of the atmosphere causes red to turn to blue, which is especially noticeable when the star is rising or setting, and thus near the horizon.

A region of the sky to explore: the constellation of Orion

Orion is certainly one of the most spectacular constellations. Situated on the equator, it is visible from all points of the Earth, but not all year round. In most cultures, Orion is regarded as a hunter but, for the Greeks, he also possessed the gift of being able to walk on water. Observed from the north coast of the Mediterranean, when the constellation passes the meridian, Orion does indeed seem to be just above the sea. Loved by Eos, goddess of the Dawn, who became Aurora for the Romans, Orion was seduced by her one evening in mid-May – at this time, the constellation vanishes from sight over the horizon. When Orion disappeared in the light of the setting Sun, Eos wept, and her tears became the dew. These amorous doings of the gods could not remain a secret, probably because of Apollo's indiscretions, and Eos was therefore overcome with shame on her return in the early morning, which explained the reddening of the sky at dawn.

Like all hunters, Orion is followed by dogs, including one very important one, symbolized by Sirius, the most brilliant star in the sky. In Ancient Egypt, Sirius, which was easy to see because of its exceptional brightness, began to appear in the early morning, when the heat was becoming overwhelming. His arrival was a sign that the waters of the Nile were rising and the agricultural year was beginning. But the Egyptian calendar had exactly 365 days. Over a few years, the religious festivals slowly shifted in relation to the astronomical and agricultural reality (the year is 365 days and just under 6 hours) and no longer corresponded to the activities they were supposed to celebrate: by the time 730 years had passed, there was a gap of six months, and priests were celebrating the Nile flood in the middle of the drought! From that time onwards, it became natural to read the seasonal calendar from the astronomical signs in the sky. From this practice was born the proverb: "The sky never lies!"

You should practice distinguishing the colors of the stars in Orion with your naked eye. To identify the two main stars it is useful to remember that, contrary to what their initials imply, Rigel is blue and Betelgeuse is red!

Betelgeuse (or **Alpha** (α) **Orionis**), from Beit Algueze or Bed Elgueze, probably deriving from the Arabic *ibt al jauzah*, "the giant's arm," is 520 light years away from the Sun. It is the only spectacular variable among the very big stars. The variability period is 5.7 years, and it is now, in 2004, at the maximum of its luminosity. Betelgeuse was the first star for which interferometry was used to measure its diameter, in 1920. It is truly enormous. During the course of its variation cycle, this diameter goes from 550 to 920 times that of the Sun, and its intrinsic luminosity oscillates between 7,600 and 14,000 times that of the Sun.

16 Thursday
- Maximum of the Piscids meteor shower (8 meteors an hour)

17 Friday

18 Saturday
- 11:40 a.m. First quarter of the Moon
- 6:06 p.m. Minimum of Algol

19 Sunday

20 Monday

21 Tuesday
- 7:43 a.m. Winter solstice

22 Wednesday

23 Thursday

24 Friday

25 Saturday

26 Sunday
- 10:07 a.m. Full Moon

27 Monday
- 2 p.m. Moon passes the apogee: 406,488 km

28 Tuesday
- The Moon and Saturn move closer together. Visible all night (minimum 5.1° at 12:42 a.m.)

29 Wednesday
- 3 p.m. Greatest western elongation of Mercury (22.4°)

30 Thursday

31 Friday

The mass of Betelgeuse is about 20 times that of the Sun, and its density is less than a ten-thousandth that of air. Look closely at its color, which is characteristic of type M supergiants and corresponds to a surface temperature of 3,100 K.

Rigel (or **Beta** (β) **Orionis**), from *rijl al jauzah al yusra*, "the giant's left leg," is another supergiant and seventh in order of the most brilliant stars. Situated 900 light years away, it reveals a surface temperature of 12,000 K. Its diameter is 50 times greater than the Sun's, and its luminosity 57,000 times greater. Look also at its color, which is that of a blue supergiant. You will find, 9" away, a relatively luminous companion with a magnitude of 6.7. Rigel's mass is of the order of 50 solar masses, which is like a rapidly evolving star, with a semi-regular variation of just under 29 days seeming to indicate slight pulsations.

Mintaka, **Alnilam** and **Alnitak** (or **Delta** (δ) **Epsilon** (ε) and **Zeta** (ζ) **Orionis**), the three stars of the belt, form one of Orion's most well-known features. Depending on the country, they are known as the Line, the Golden Grains, the String of Pearls or the Three Arrows. Look at Mintaka, a beautiful blue giant 5,000 times more luminous than the Sun, with a surface temperature of 28,000 K.

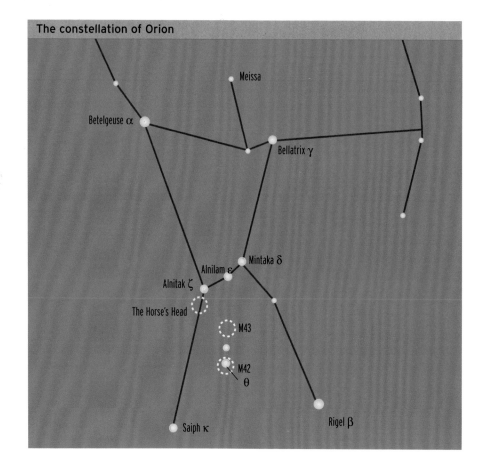

The constellation of Orion

Meissa

Betelgeuse α

Bellatrix γ

Mintaka δ

Alnilam ε

Alnitak ζ

The Horse's Head

M43

M42
θ

Saiph κ

Rigel β

THE SKY IN DECEMBER

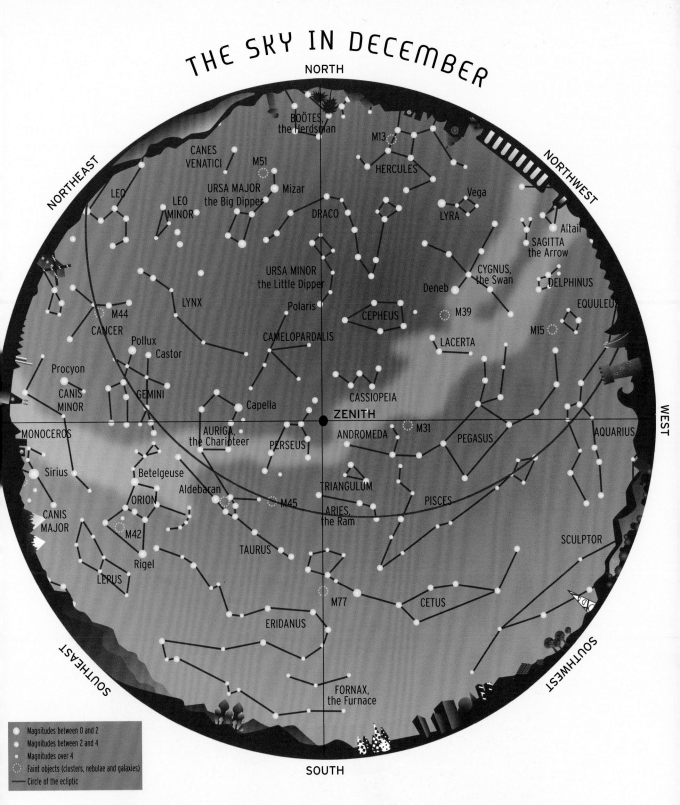

NORTH

BOÖTES,
the Herdsman

M13

HERCULES

CANES
VENATICI   M51

LEO   URSA MAJOR   Mizar   Vega
LEO   the Big Dipper   DRACO   LYRA   Altair
MINOR   SAGITTA
the Arrow

CYGNUS,
the Swan   DELPHINUS

URSA MINOR
LYNX   the Little Dipper   Deneb   M39   EQUULEUS

Polaris   CEPHEUS   M15

M44   CAMELOPARDALIS   LACERTA

CANCER
Pollux
Castor

Procyon   GEMINI   CASSIOPEIA

CANIS   ZENITH
MINOR   Capella   M31
AURIGA,   ANDROMEDA   PEGASUS   AQUARIUS
MONOCEROS   the Charioteer   PERSEUS

Sirius   Betelgeuse   TRIANGULUM   PISCES
Aldebaran   M45
CANIS   ORION   ARIES,   SCULPTOR
MAJOR   M42   the Ram
Rigel   TAURUS
LEPUS   M77   CETUS

ERIDANUS

FORNAX,
the Furnace

NORTHEAST   **NORTHWEST**

WEST

SOUTHEAST   **SOUTHWEST**

SOUTH

- ⊙ Magnitudes between 0 and 2
- ● Magnitudes between 2 and 4
- • Magnitudes over 4
- ⊙ Faint objects (clusters, nebulae and galaxies)
- — Circle of the ecliptic

How to use this chart

Hold the chart above your head, matching the word SOUTH that appears at the edge of it with the geographical south of the place you are observing from. Use a compass to help you do this.

This chart shows the sky that is visible at a latitude of 45°

If you are further north or further south, Polaris will be higher or lower in the sky.

Chart of the sky visible at 10 p.m. EST

at the beginning of the month; at 9 p.m. EST in the middle of the month; at 8 p.m. EST at the end of the month.

A nursery of stars for you to observe

M42, the great Orion Nebula, reveals different charms according to which instruments you observe it with. On a beautiful moonless night, you can make out a slight nebulosity with the naked eye: Orion's sword and baldric form a T-shape, the vertical line being the sword, with M42 at its center. Through binoculars, the spectacle is dazzling and you will be able to spot a large butterfly-shaped nebulosity. With a telescope of about 8 inches (200 mm) fitted with a wide-angle eyepiece and with low magnification, M42 appears in all its splendor. With at least this diameter, you can begin to make out a greenish color. Notice the small group of stars in the center – this is **Theta** (τ) **Orionis**, a multiple system also known as the Trapezium or the Quadruple Star. The nebula is about 1,700 light years away. This is a very complex region of star formation, but

M42
The central part of the Orion Nebula is observed here in the near infrared.

Horsehead Nebula
This spectacular nebula is found in the constellation of Orion, in the vicinity of the star Alnitak.

the chemical composition of the interstellar material is relatively simple: for 1 billion atoms of hydrogen, there are 100 million atoms of helium, 1 million atoms of carbon, 300,000 atoms of oxygen, 200,000 atoms of nitrogen, and traces of sulfur, neon and chlorine. Let your binoculars or telescope wander over this beautiful region of the sky. You will discover fine draperies, stars gleaming like jewels and dark masses of interstellar dust, but the most remarkable objects are not visible to the naked eye. The striking Horsehead Nebula with its evocative shape, a large cloud near Alnitak, is only revealed by means of photography. Here, stars are even now in the process of formation.

Practical astronomy

Optical instrument of the month: the Schmidt–Cassegrain telescope

Drawing of a Schmidt–Cassegrain telescope
This type of telescope is generally fitted, as here, with a fork mounting.

(labels on drawing: Telescopic sight, Tube, Eyepiece, Fork mounting)

Although it is traditional to start off with a small Newtonian telescope, generally of the 115/900 type, when you want to use a larger diameter to make better observations or start taking photographs, the Schmidt–Cassegrai telescope is the one people generally go for, and it is very popular among amateur astronomers. The most common ones are 8 inches (200 mm) in diameter, which is a clear progression from the traditional 4 inches (100 mm), since the amount of light collected is three times greater.

A compact instrument

The basic principle consists in using a spherical mirror instead of a parabolic mirror. Spherical mirrors are easier to manufacture, and therefore less expensive. But they give images that are affected by "spherical aberration," a slight blur that is a nuisance when stargazing. However, this haziness is corrected by the very thin correcting lens, called the "Schmidt plate" in homage to its inventor, Bernard Schmidt, who created the first combination of this type in 1930.

The main advantage of Schmidt–Cassegrain telescopes is their compactness, which explains their success: a large diameter becomes easily transportable and offers an aperture ratio that is sufficient to photograph faint objects.

The mechanics and setting into position are very simple. Schmidt–Cassegrains generally come with motors, they are quick to get working, and are comfortable to use. Photographic equipment can easily be adapted to them. Moreover, the observer stands behind the instrument, as with a spyglass, whereas with a Newtonian telescope you have to place yourself at its open end. It is therefore easier to point with a Schmidt–Cassegrain telescope.

With these instruments, minimum magnification remains equal to D/7, the diameter D being expressed in millimeters. Certain tasks, on the other hand, like measuring double stars, demand strong magnification; but there is no point, even under good observation conditions, in magnifying 1,000 times without very good quality equipment and an excellent site.

Accessories

Nearly all eyepieces are 1.25 inches (31.75 mm) in diameter, now a standard measurement. These telescopes usually come with an angled eyepiece which makes the observer's task easier and saves getting a stiff neck.

Optical diagram of the Schmidt–Cassegrain telescope
The passage of the light reveals the extreme compactness of this type of instrument.

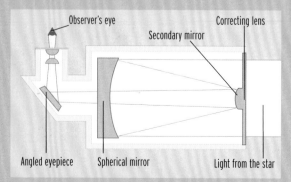

(labels on diagram: Observer's eye, Secondary mirror, Correcting lens, Angled eyepiece, Spherical mirror, Light from the star)

However, the telescopic sights on Schmidt–Cassegrains can be a weak point, and they may lack guides, which are ideal for photography. The sights are usually small spyglasses of average quality that are not always easy to adjust parallel to the lens and ensure that they remain fixed. The sights without a lens, of the Starpointer type, which are easy to use and adjust, are becoming more and more popular. Schmidt–Cassegrains are usually fitted with fork mountings, which combine compactness with ease of use and transportation.

Prices

You will find shops selling Schmidt–Cassegrains of 5 to 14 inches (125 to 355 mm). Prices are obviously dependent on diameter. The smallest telescopes can cost less than $1,000, against more than $10,000 for the largest ones. These prices include the mechanized mounting, and certain large diameter models even include an automatic pointing system of the "Go-To" variety.

What should you observe?

Everything is accessible to a Schmidt-Cassegrain telescope. You will get a great deal of pleasure from observing the Moon, being careful to add an eyepiece filter so as not to be too hampered by the quantity of light being received. As for the planets, the 8-inch (200 mm) diameter enables you to make drawings in some comfort.

But it is for faint objects that Schmidt-Cassegrains are the most useful; they are usually fitted with good quality graduated circles, which allow you to point them from their coordinates. It is therefore useful to have equipment available with which to calculate the positions you want to reach: either a computer that can determine the observation conditions, or a clock that is able to keep sidereal time, as well as a sky atlas and ephemerides for the bodies of the solar system. With instruments from 8 inches (200 mm) in diameter, you can take good photometric measurements and carry out regular programs of observation on variable stars, asteroids and comets.

Jupiter
The giant planet and three of its satellites through a 6-inch (150 mm) Schmidt-Cassegrain.

The Moon
The Mare Crisium is seen through a 12-inch (300 mm) Schmidt-Cassegrain telescope.

Venus
This crescent of Venus is observed with an 8-inch (200 mm) Schmidt-Cassegrain.

Stars in the process of formation
This cluster contains about 50 massive hot stars and is surrounded by a multitude of cold young stars in the process of formation.

Discovering star clusters

Star groupings like the Pleiades, the Hyades, the Perseus clusters or the Beehive cluster are called "open clusters" or "galactic clusters." They are stars born at the same time, in the same cloud of interstellar material, but evolving differently, in accordance with their mass.

In addition to the few renowned clusters that are visible to the naked eye, our galaxy includes a large number of these groupings. More than 1,000 are listed but some of them are questionable, several are not real associations of stars linked by gravitation and only a few hundred have distances that are known, and these are being carefully researched. The Magellanic Clouds and the galaxies nearest to our own possess 4,200 and 2,000 of these clusters respectively, and they are the subjects of close study.

Each of the clusters is characterized by its age, which astronomers determine by recording the state of evolution of the most massive stars. Most of the clusters listed are less than 500 million years old. About 20, however, are nearly a billion years old. The average mass of these clusters varies between 350 and 7,000 solar masses. There is a distinction between rich clusters and classical clusters. Rich clusters, like M11 in the constellation of Scutum, have a large

The Jewel Box
This beautiful galactic cluster in the southern hemisphere has been given an evocative name.

central concentration, and average distances between the stars can approach those between the planets of the solar system; they are very strongly linked by gravitation and take an enormous amount of time to disperse. Clusters with a low mass, on the other hand, see their stars escape quite rapidly. In a few tens of millions of years, they lose more than half their numbers. Typical clusters disperse in a few hundred million years and so manage to go round the galaxy several times before disappearing, leaving the stars they were sheltering to move as they please. They revolve around the galaxy in about 220 million years.

Open clusters are precious objects that enable us to further our knowledge of the stars' evolution, without having to measure how far away they are. When stars in the same cluster are classified according to their color and luminosity, a diagram can be made, called an HR diagram, in memory of the two astronomers Ejnar Hertzsprung and Henry Russell, who were the first to undertake this type of classification at the beginning of the 20th century. Stars begin their entire evolution in the same way, in a huge cloud mainly containing hydrogen, helium and traces of heavier elements like carbon, nitrogen, sulfur, etc. There are about 16 times more hydrogen than helium atoms in structures like these with very low density, about a billion billion times less than that of our atmosphere. When

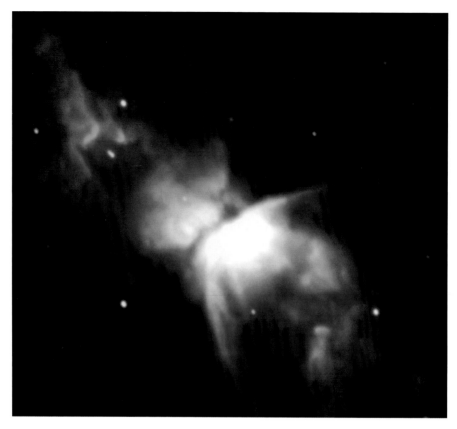

The Butterfly Nebula
The explosion of a medium-sized star generated this nebula.

In this book, temperatures are given in degrees Kelvin, or K; we are then talking about absolute temperatures. Absolute zero corresponds to matter that is in a totally immobile state and, expressed on a temperature scale of centigrade, this would be -273.15°C. On the Kelvin scale, water boils at 373.15 K and solidifies at 273.15 K.

such a cloud is disturbed by an outside accident, for example a star exploding or a density wave passing and thus causing contraction, several hundred concentrations cause the temperature to rise. This is accentuated by the ambient dust, which stops heat leaving the stellar cocoon that has thus been made. When a concentration reaches the size of our solar system, it becomes opaque and its temperature reaches several hundred degrees Kelvin. A first catastrophe then occurs: the dihydrogen molecules dissociate and certain atoms ionize, with their electrons being pulled out. The concentration starts to diffuse energy outwards and becomes luminous. The collapse takes the form of a cataclysm, which considerably increases the pressure of the gases, and therefore the temperature. The process ceases when equilibrium is reached between the gravitation and radiation pressure. The protostellar cloud is then about 100 times bigger than the Sun. This rapid phase (a few months) is known as "the T Tauri phase," from the name of the star in the constellation of Taurus, whose breakneck speed of change was studied in 1942. Stars of this type show great variations in brightness, and their spectrum is rich in lines of emission whose structure gives a picture of the star's collapse. The surface temperature

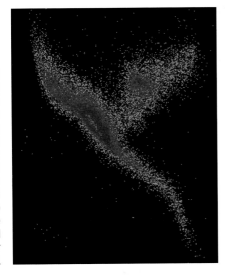

Color/ luminosity diagram
From measurements taken by the Hipparcos satellite, 40,000 stars have been classified according to their magnitude (vertical) and color (horizontal).

is some 4000 K, the energy radiated being 100 times greater than that of the Sun. At this stage, a slower contraction begins and the star becomes less luminous, the core temperature slowly increases and rotation speeds up. The material that is in close proximity and subject to universal gravitation spreads out in a great ring that revolves around the central concentration. It is in this ring that the planets are formed, through secondary contractions, like kinds of lumps appearing in this primitive "soup" which will become a stellar system.

When the temperature of the core contraction reaches 15 million degrees Kelvin, the hydrogen is transformed into helium and diffuses energy by thermonuclear fusion reaction, and the star lights up. This phase can be more or less long, depending on the star's mass. For a star like the Sun, it lasts about 50 million years; for stars of 0.5 solar masses it requires 100 million years, as against 100,000 years for a star of 10 solar masses. Several hundreds of stars are

born in this way at the same time. The most massive ones consume their hydrogen much more quickly than the others. For the whole duration of this nuclear combustion, the star is living its adult life.

As an adult, a star appears in the part of the HR diagram known as "main sequence." Then, once the hydrogen in the core has been exhausted, its diameter and luminosity increase, and the star becomes a "red giant" and leaves the main sequence. We know that, in a cluster, stars are all born at the same time and the first to leave are the most massive ones. By noting the place where the main sequence ends, we can learn the age of the most massive stars, and therefore the age of the cluster.

M68
This globular cluster is situated in the constellation of Hydra.

The new star catalogs

Sky chart
Detail of an advanced chart showing stars and nebulae accessible to small optical instruments.

For a long time, stars were thought to be fixed, but we now know that they are moving in relation to one another. It has therefore been vital for us to improve our knowledge of their positions, and the measuring of this is done by "astrometry." Catalogs of stars have been compiled since Classical times, the first one having been made by Hipparchus, in the 2nd century B.C. At that time, positions were noted to within the nearest few minutes of arc, but the precision of these catalogs has slowly developed over the course of history. In the 17th century, the invention of spyglasses and micrometers enabled measurements to be improved, until they were precise to within a fraction of a second of arc in the 20th century.

However, the positions of celestial objects are prone to several "imperfections" which cause them to change. Firstly, the nearest stars are affected by parallax (see p. 34), which makes their position oscillate over the course of the year in accordance with the Earth's movement around the Sun. Secondly, they move among their neighbors according to the various motions affecting the Galaxy. So that we may know what these motions are, they are separated into two elements, following two different directions, namely their movement in space relative to the sun, or "proper motion," and their projection along the line of sight, or "radial velocity." A star may be said to be known "astrometrically" when its angular position in the sky, its distance, and its proper motion have been precisely determined.

The Hipparcos and Tycho catalogs
The measurements of positions carried out on the ground were, until the arrival of the astrometric space programs, precise to about a tenth of a second of arc, that is to say the angle equivalent to the size of a man seen from 3,000 kilometers away. The Hipparcos satellite was launched in 1989 and operated until 1993, and enabled the positions of stars to be measured from space, where the images of these stars are not disturbed by the Earth's atmosphere. Thus, in 1996 a catalog appeared which gathered together 120,000 stars with an astronomical precision superior to one millisecond of arc – in other words the size

The ambitions of GAIA

The GAIA mission, like Hipparcos a European Space Agency program, is in the process of development. The launch of this satellite will take place in 2009. The objective is to create, from space, a catalog that is a thousand times more precise than that of Hipparcos. The number of stars involved will be ten times greater. Thanks to GAIA, it will be possible to determine the proper motions of stars belonging to the Magellanic Clouds and the dynamics of galaxies in the Local Group. GAIA will radically change our understanding of the Universe. With the help of this program, we should discover 50,000 new planetary systems, 100 million pairs of stars, between 100,000 and 1 million new bodies in the solar system, tens of thousands of cold stars and, outside our galaxy, thousands of supernovae and quasars. The data-gathering program will last five years, and several more years will be required to go through it all and publish the catalogs.

Field of stars
Thanks to Hipparcos, the position and distance of one million stars have been calculated with unparalleled precision.

The Hipparcos satellite

of a man seen on the Moon! A second catalog followed, called Tycho in homage to Tycho Brahe who, at the end of the 16th century, carried out the best measurements of positions ever obtained without optical help. It lists one million stars observed by Hipparcos with, this time, a precision of 20 to 30 milliseconds of arc, i.e., about a tenfold improvement on the previous situation. Better knowledge of these parameters is of enormous importance. The number of stars whose distance is well known has been multiplied by more than 100. This has enabled a considerable improvement to be made in the precision of classifications and in the HR diagram, which is used to determine the outline of stellar evolution. The consequences for astrophysics are enormous, and have encouraged astronomers to launch the idea of a super Hipparcos, a new program called GAIA.

General sky chart for the
Northern Hemisphere

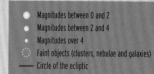

- Magnitudes between 0 and 2
- Magnitudes between 2 and 4
- Magnitudes over 4
- Faint objects (clusters, nebulae and galaxies)
— Circle of the ecliptic

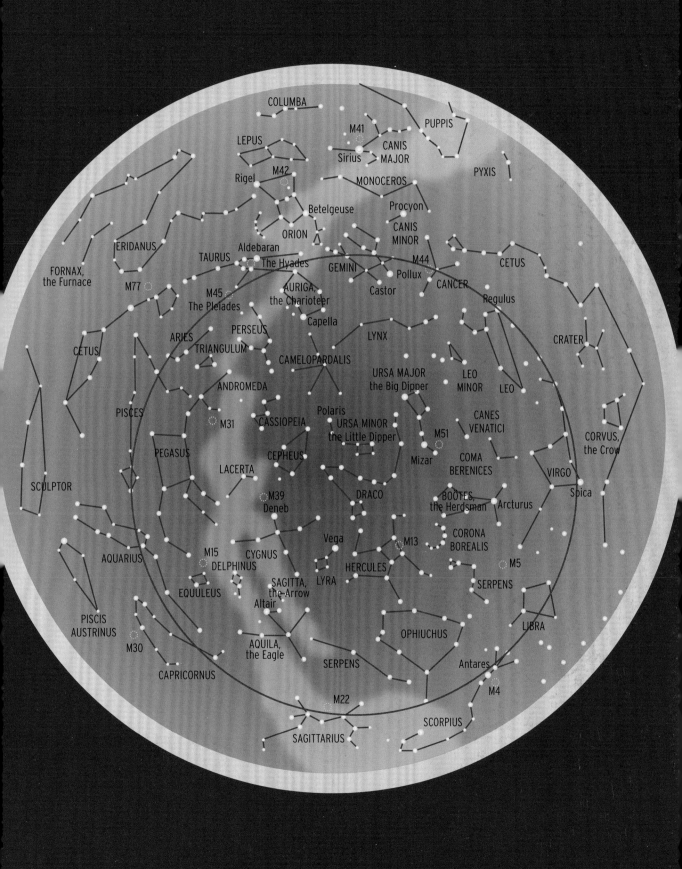

Practical information

Further reading

■ Books

Advanced Skywatching (a.k.a. *Backyard Astronomy* in softcover) by Robert Burnham, Alan Dyer, Robert Garfinkle, Martin George and Jeff Kanipe (Nature Company/Time-Life: San Francisco, 1997).

Astronomical Centers of the World by Kevin Krisciunas (Cambridge University Press: Cambridge, 1988).

Astronomy: The Definitive Guide by Robert Burnham, Alan Dyer, and Jeff Kanipe (Weldon-Owen: Sydney, 2002).

The Backyard Astronomer's Guide: revised editon by Terence Dickinson and Alan Dyer (Firefly Books: Toronto, 2001).

The Beginner's Observing Guide by Leo Enright (Royal Astronomical Society of Canada: Toronto, 1999).

Cambridge Encyclopedia of Meteorites by O. Richard Norton and Alan E. Rubin (Cambridge University Press: Cambridge, 2002).

Comets: Creators and Destroyers by David H. Levy (Simon and Schuster: New York, 1998).

The Compact NASA Atlas of the Solar System by Ronald Greeley and Raymond Batson (Cambridge University Press: Cambridge, 1997).

David Levy's Guide to the Night Sky by David H. Levy (Cambridge University Press: Cambridge, 2001).

Eclipse by Duncan Steel (Headline: London, 1999).

Exploring the Sky by Day by Terence Dickinson (Firefly Books: Toronto, 1988).

Eyewitness Handbooks: Stars and Planets by Ian Ridpath (Dorling Kindersley, London: 1998).

Glorious Eclipses by Serge Brunier and Jean-Pierre Luminet (Cambridge University Press: Cambridge, 2000).

Great Comets by Robert Burnham (Cambridge University Press: Cambridge, 2000).

The Magnificent Universe by Ken Crosswell (Simon and Schuster: New York, 1999).

Mars: The Lure of the Red Planet by William Sheehan and Stephen James O'Meara (Prometheus Books: New York, 2001).

The New Solar System, Beatty, Peterson and Chaikin, ed. (Sky Publishing/Cambridge University Press: Cambridge, 1999).

Nightwatch by Terence Dickinson (Firefly Books: Toronto, Ontario, 1998).

The Night Sky Observer's Guide (two volumes) by George Robert Kepple and Glen W. Sanner (Willmann-Bell: Richmond, VA, 1998).

Observing Meteors, Comets, Supernovae, and other Transient Phenomena by Neil Bone (Springer-Verlag: London, 1998).

Observing the Moon: The Modern Astronomer's Guide by Gerald North (Cambridge University Press: Cambridge, 2000).

The Planet Observer's Handbook by Fred Price and John Westfall (Cambridge University Press: Cambridge, 2000).

Seeing in the Dark by Timothy Ferris (Simon and Schuster: New York, 2002).

Skywatching by David Levy (Nature Company/Time-Life: San Francisco, 1994).

The Southern Sky Guide by David Ellyard and Wil Tirion (Cambridge University Press: Cambridge, 2001).

Star and Sky: Discovery Travel Adventures; Robert Burnham, ed. (Discovery Communications: London, 2000).

Star Hopping for Backyard Astronomers by Alan M. MacRobert (Sky Publishing: Cambridge, Massachusetts, 1993).

Star Hopping: Your Visa to Viewing the Universe by Robert Garfinkle (Cambridge University Press: Cambridge, 1994).

Stars and Planets: A Viewer's Guide by Gunter Roth (Sterling Publishing: New York, 1998).

Summer Stargazing by Terence Dickinson (Firefly Books: Toronto, Ontario, 1996).

Turn Left at Orion by Guy Consolmagno and Dan M. Davis (Cambridge University Press: Cambridge, 2000).

The Universe and Beyond by Terence Dickinson (Firefly Books: Toronto, Ontario, 1996).

The Universe and How to See It by Giles Sparrow (Reader's Digest Books: New York, 2001).

A Walk Through the Southern Sky by Milton D. Heifetz and Wil Tirion (Cambridge University Press: Cambridge, 2000).

■ Atlases

Atlas of the Lunar Terminator by John E. Westfall
(Cambridge University Press: Cambridge, 2000).
Bright Star Atlas by Wil Tirion (Willmann-Bell:
Richmond, VA, revised 2002).
Cambridge Star Atlas by Wil Tirion (Cambridge
University Press: Cambridge, 1996).
Great Atlas of the Stars by Serge Brunier and Akira Fuji
(Firefly Books: Toronto, Ontario, 2001).
The Hatfield Photographic Lunar Atlas, Jeremy Cook,
ed. (Springer-Verlag: London, 1999).
Millennium Star Atlas by Roger W. Sinnott and
Michael A.C. Perryman (Sky Publishing/European
Space Agency: Cambridge, 1997).
SkyAtlas 2000.0 by Wil Tirion and Roger W. Sinnott
(Sky Publishing: Cambridge, MA, 1998).
Uranometria 2000.0 Deep Sky Field Guide by Wil Tirion,
B. Rappaport, and G. Lovi. (Willmann-Bell:
Richmond, VA, revised 2001).

■ Magazines and Journals

Amateur Astronomy
To subscribe call: (352) 490-9101
www.amateurastronomy.com
Astronomy
To subscribe call: (800) 533-6644
www.astronomy.com
*The Journal of the Royal Astronomical Society of
Canada*
To subscribe call: (416) 924-7973
www.rasc.ca
Mercury
To subscribe call: (415) 337-1100
www.astrosociety.org
Planetary Report
To subscribe call: (626) 793-5100
planetary.org
Star Date
To subscribe call: (512) 471-5285
www.stardate.org
SkyNews
To subscribe call: (866) 759-0005
www.skynewsmagazine.com
Sky and Telescope Magazine
To subscribe call: (800) 253-0245
www.skyandtelescope.com

Websites

www.backyardastronomy.com
A companion website to Dickinson and Dyer's
popular book *The Backyard Astronomer's Guide.*
www.seds.org/
A website called The Nine Planets which features a
fascinating multimedia tour of the solar system.
www.nationalgeographic.com/solarsystem
A 3-D virtual reality tour of the sun, planets and
extraterrestrial weather patterns throughout the
solar system.
cdsweb.u-strasbg.fr/~heck/sfworlds.htm
A searchable database called StarWorlds of more
than 11,000 astronomy websites.
astronomylinks.com
A commercial site of astronomy links, which are
organized by category.
www.cv.nrao.edu/fits/www/astronomy.html
AstroWeb is a comprehensive list of almost every
astronomy site on the Internet.

■ Images

pds.jpl.nasa.gov/planets
A website called Welcome to the Planets featuring a
comprehensive collection of images from NASA's
planetary exploration program.
www.astronomy.ca/images/
The Astronomy Image Library featuring a searchable
database of all astronomical images on the web.
antwrp.gsfc.nasa.gov/apod/astropix.html
The Astronomy Picture of the Day website.
www.jpl.nasa.gov/galileo
All the latest news and images from the Galileo
probe during its current journey to Jupiter.

www.hubblesite.org

This website contains links to the latest publicly released images from the Hubble Space Telescope.

■ Educational

www.astronomycafe.net

Sten Odenwald's Astronomy Café provides concise, cogent answers to all the frequently and not so frequently asked questions about space and astronomy.

space.jpl.nasa.gov

A solar system simulator created by the Jet Propulsion Laboratory at NASA.

solar-center.stanford.edu

A website that provides students with a collection of on-line educational activities about the sun.

■ Special Interest

mpfwww.jpl.nasa.gov

The website of NASA's Mars Exploration Program containing the latest news, information and multimedia images.

www.lpi.usra.edu/lpi.html

A website by the Lunar and Planetary Institute featuring all you need to know about the moon and its exploration by astronauts.

www.earth.nasa.gov

The homepage of NASA's Earth Science Enterprise: a project dedicated to understanding the Earth and the effects of natural and human-induced changes on the global environment.

spaceflight.nasa.gov/station

A website containing news, images and information from the astronauts currently working on the ISS (International Space Station).

saturn.jpl.nasa.gov/index.cfm

Full coverage of the Cassini space probe, which is currently on its way to explore Saturn and its moon Titan.

encke.jpl.nasa.gov

A website containing information and images of comets and meteors as supplied by NASA's Jet Propulsion Laboratory.

nssdc.gsfc.nasa.gov/planetary/giotto.html

Images and information about the Giotto space mission and its studies of the comets Halley and Grigg-Skjellerup.

www.seti.org

The SETI (Search for Extra-Terrestrial Intelligence) website. The project's mission is to search for evidence of extraterrestrial life in the universe.

sohowww.nascom.nasa.gov

Images and data obtained by the SOHO (Solar and Heliospheric Observatory) space probe.

www.bbso.njit.edu

A site containing studies and images of the Sun as observed from the Big Bear Solar Observatory in California.

■ Astronomy Associations and Societies

Amateur Astronomer's Association
www.aaa.org/
American Association of Astronomers
www.corvus.com/
American Astronomical Society
dda.harvard.edu/
American Meteor Society
www.amsmeteors.org/
Astronomical League
www.astroleague.org
Astronomical Society of the Pacific
www.astrosociety.org
Planetary Society
planetary.org
Royal Astronomical Society of Canada
www.rasc.ca

■ Amateur Observing Associations

American Association of Variable Star Observers
www.aavso.org
Association of Lunar and Planetary Observers
www.lpl.arizona.edu/alpo/
International Occultation Timing Organization
www.lunar-occultations.com/iota/iotandx.htm

Planetariums

Adler Planetarium
Chicago (312) 922-STAR
www.adlerplanetarium.org
Albert Einstein Planetarium
Washington (202) 357-1400
www.nasm.edu/nasm/planetarium/Einstein.html

Aldrin Planetarium
Florida (561) 832-1988
www.sfsm.org
Christa McAuliffe Planetarium
New Hampshire (603) 271-7827
www.starhop.com/home.htm
Dreyfuss Planetarium
Newark (973) 596-6529
www.newarkmuseum.org/planetarium
Fiske Planetarium
Colorado (303) 492-5001
www.colorado.edu/fiske
Griffith Observatory Planetarium
Los Angeles (323) 664-1191
www.griffithobs.org/Planetarium.html
Hayden Planetarium
New York (212) 769-5200
www.amnh.org/rose/haydenplanetarium.html
Halifax Planetarium
Halifax (902) 424-7353
halifax.rasc.ca/hp
Henry Buhl, Jr., Planetarium
Pittsburg (412) 237-3400
www.carnegiesciencecenter.org/exhibits/planet.asp
Manitoba Planetarium
Winnipeg (204) 956-2830
www.manitobamuseum.mb.ca/pl_info.html
Macmillan Planetarium
Vancouver (604) 738-7827
www.hrmacmillanspacecentre.com
Miami Museum of Science Planetarium
Miami (305) 646-4200
www.miamisci.org/www/eventsplan.html
Montreal Planetarium
Montreal (514) 872-4530
www.planetarium.montreal.qc.ca
Noble Planetarium
Fort Worth (817) 255-9300
fortworthmuseum.org/noble.html
Roberta Bondar Earth and Space Center
Planetarium
Toronto (416) 491-5050 ex 2227
www.senecac.on.ca/bondar/index.html
William Knox Planetarium
California (510) 642-5132
www.lhs.berkeley.edu/planetarium

■ **Observatories**
Apache Point Observatory

New Mexico
www.apo.nmsu.edu
Big Bear Solar Observatory
California
www.bbso.njit.edu
Caltech Submillimeter Observatory
Hawaii
www.submm.caltech.edu/cso
Carnegie Observatories
California
www.ociw.edu
Cincinnati Observatory Center
Ohio
www.cincinnatiobservatory.org
David Dunlop Observatory
Ontario, Canada
ddo.astro.utoronto.ca/ddohome
Dominion Astrophysical Observatory
British Columbia, Canada
www.hia-iha.nrc-cnrc.gc.ca/dao/index_e.html
Kitt Peak National Observatory
Arizona
www.noao.edu/kpno
McDonald Observatory
Texas
www.as.utexas.edu/mcdonald
Millstone Hill Observatory
Massachusetts
hyperion.haystack.edu
Mount Wilson Observatory
California
www.mtwilson.edu

Appendices

■ **Angular momentum** Measurement of the amount of rotation. This amount depends both on angular speed and on the size of the bodies in rotation. The larger the body and the faster it turns, the higher its angular momentum. In an isolated system (not subject to influence from outside), the angular momentum is invariable: if the system revolves at less and less speed, its size increases; if its size decreases, it speeds up. This is clearly illustrated by the ice-skater who turns more and more quickly when he draws his arms in close to him, and slows down when he stretches them out.

■ **Angular separation** Angle separating two stars.

■ **Annular eclipse of the Sun** A central eclipse, in other words, when the Moon passes right in front of the Sun, but in this case, because the Moon is near its apogee and so at maximum distance from the Earth, it appears smaller and does not mask the Sun completely.

■ **Anomalistic period** Time taken by a star to return to the periastron of its orbit.

■ **Aphelion** The point in the Moon's or a star's orbit that is farthest from the Sun.

■ **Apogee** The point in the Moon's or a star's orbit that is farthest from the Earth.

■ **Asterism** A group of stars.

■ **Asteroid** A minor planet of the solar system, whose dimensions do not exceed a few hundred kilometers. Most asteroids are small, and their trajectories lie between the orbits of Mars and Jupiter.

■ **Astronomical refraction** Change in the apparent position of a star when its light has passed through the Earth's atmosphere. At the horizon, this refraction lifts the images of all the stars by half a degree. At the zenith, it has no effect.

■ **Astronomical unit (AU)** Unit of measurement, being the average distance between the Sun and the Earth, in other words 149,597,870 kilometers.

■ **Black hole** Extreme gravitational phenomenon from which light cannot escape. At the end of their evolution, some very massive stars may become black holes.

■ **Bolide** A luminous ball produced by a meteorite weighing several grams, or even kilograms, striking the Earth's atmosphere.

■ **Celestial equator** Apparent course of the Sun at the time of an equinox.

■ **Chromatic aberration** Iridescence from images caused by the lens of a telescope or other optical instrument. In a simple lens, the objective acts as a prism and scatters the colors that compose the light in different ways. You then get superimposed images of different color and size.

■ **Chromatic scintillation** Phenomenon produced by the Earth's atmosphere. Acting as a prism, the atmosphere disperses the light from the stars. When the star is very luminous, like Sirius, there is enough energy in the scattered colors for these to be visible. The turbulence in the atmosphere causes it to turn from red to blue. This is especially noticeable when the star is rising or setting and therefore close to the horizon.

■ **Cluster** Group of stars linked by gravitation. A cluster is called "globular" when the number and mass of stars is large enough for the whole group to stay together in a stable group. It is called "open" if this is not the case, and stars will leave the cluster within a few tens of millions of years.

■ **Comet** Fossil remains left over from the formation of the solar system, with a nucleus of dust and water.

Glossary

Dimensions vary between a few kilometers and a few tens of kilometers. Comets have very long orbits and are only visible when they are approaching the central region of the solar system. They then partially melt, leaving behind them a tail of gas and dust that can reach several hundreds of millions of kilometers.

■ **Conjunction** When two solar system bodies appear to be close together.

■ **Constellation** Artificial grouping of stars forming the basis of images and legends.

■ **Doppler-Fizeau (effect)** Change in the color of light due to movement of the light source. If the source is moving toward the observer, the light becomes bluer; if receding, the light becomes redder. If the light is broken down into a spectrum, all the spectral lines shift toward the blue end of the spectrum when the source is moving toward the observer, and toward the red end of the spectrum when it is receding.

■ **Double star** Two stars linked by universal gravitation.

■ **Draconic period** Time taken by a star to return to the same node of its orbit.

■ **Dwarf (star)** Star with low luminosity.

■ **Earthlight** or **Earthshine** Sunlight reflected from the Earth onto the Moon. When we can only see a thin crescent from Earth, the Sun is mainly illuminating the side of the Moon that is hidden from view. The Sun is then in the same half of the sky as the Moon, casting its light directly onto the face of the Earth. The Earth reflects this light back into the sky, illuminating the Moon's surface just as the full Moon sheds light on the surface of the Earth.

■ **Ecliptic** Plane which passes through the center of the Sun and contains the Earth's orbit.

■ **Ellipticity** The degree of deviation of an ellipse from a circle.

■ **Elongation** Apparent distance, usually of a planet from the Sun.

■ **Equinox** Time of year when the Sun's apparent trajectory crosses the celestial equator, meaning that day and night are of equal length.

■ **Escape velocity** Speed which needs to be reached to escape from a star's gravitational field.

■ **Galaxy** Immense mass of stars linked by gravitation. There are at least several tens of billions of stars in a galaxy.

■ **Gibbous** A word coming from the Latin *gibbosus*, "hunchback," it describes the Moon when over half of the disc we can see is illuminated. The Moon is gibbous between its first quarter and Full Moon, and between the Full Moon and the last quarter.

■ **Gravity** Force of attraction on its surroundings caused by a star's mass.

■ **Inferior planet** Planet which is always inside the Earth's orbit. There are two of these, Mercury and Venus.

■ **Kelvin (degrees)** Unit for measuring temperature. The degree Kelvin, or K, is also known as absolute temperature. Absolute zero corresponds to a state in which the particles making up matter are immobile. Expressed on the Celsius temperature scale (°C), this would be at −273.15°C. On the Kelvin scale, water boils at 373.15 K and solidifies at 273.15 K.

■ **Light year** A unit of measurement, being the distance traveled by light in a year at a speed of 300,000 km/s. Expressed in kilometers, this distance is equal to

Appendices

9,460,730.472 kilometers. The distance between the Earth and the Sun is almost 500 light seconds.

■ **Magnitude** Measurement of star brightness.

■ **Meteor** Bright streak of light visible in the sky caused by a meteroid.

■ **Meteorite** Solid fragment arriving from interplanetary space onto the surface of a planet.

■ **Meteoroid** Fragment of interplanetary matter liable to break up and produce meteorites.

■ **Meteor swarm** Cloud of dust released by comets as they pass through the central region of the solar system. When the Earth encounters these clouds, showers of shooting stars are produced.

■ **Nebula** Diffuse patch of luminosity, which may be a gas cloud or a star cloud.

■ **Node** Point at which a body's orbit intersects a reference plane, usually the plane of the ecliptic. Each month the Moon crosses the ecliptic: when it passes from south to north, the node is said to be ascending, and it is descending when it moves from north to south.

■ **Nova** Sudden increase in a star's brightness, which is linked to the end of its evolution.

■ **Occultation** An eclipse. Viewed from the Earth, the planets occasionally pass in front of distant stars. When these occultations occur, the fluctuation of stellar light gives us information about the environment on the interposing planet.

■ **Opposition** Position of a star opposite the Sun.

■ **Organic molecule** Chain of carbon atoms, including atoms of hydrogen, oxygen, nitrogen, etc.

■ **Parallax** Angular shift. If you look at close objects with each eye alternately, they appear to shift in relation to the background; in the same way, at six-monthly intervals, stars which are close appear to shift in relation to those which are farther away, because the Earth's position is changing due to its movement around the Sun. For stars which are closer, the parallax only measures a fraction of a second of arc (angle at which you can see a person 300 kilometers away).

■ **Parsec** Unit of measurement. This is the distance at which an observer should stand in order to see the Sun and Earth separated by an angle of 1° of arc. This distance is equal to 3.261 light years, or 30,857 billion kilometers.

■ **Periastron** Point when a body is closest to the star that it is orbiting.

■ **Perigee** Point in the Moon's or a star's orbit around the Earth when it is closest to the Earth's center.

■ **Perihelion** Point in the Moon's or a star's orbit around the Sun when it is closest to the Sun's center.

■ **Period of a comet** Time separating two passages of a comet past the perihelion, the point closest to the Sun. It is therefore the time taken by the body to go around the Sun once.

■ **Polar aurora** Illumination of the atmosphere caused by particles from the Sun.

■ **Pulsar** Star which is at the end of its evolution, having a very high density and revolving at high speed.

■ **Quasar** Very distant object which looks like a star and emits as much energy as a galaxy.

■ **Radial velocity** Speed at which a star moves along an optical instrument's line of sight.

Glossary

■ **Relativity** Theory of physics linking spatial coordinates with time. It states that no mass can exceed, or even attain, the speed of light. Time and distance become relative, according to the reference points from which they are observed.

■ **Resonance** Term used by physicists, more or less a synonym of synchronization. The orbits of the planets and their moons develop under the influence of their neighbors, and may end up with a synchronization known as resonance. This often takes the form of a simple relationship between two complete numbers. Thus, some asteroids are impelled to go round the Sun three times while Jupiter only goes round it twice (resonance 3/2).

■ **Retrogradation** Orbiting in a direction opposite to that of the Earth.

■ **Satellite** Body in orbital motion around a planet. The solar system has several dozen natural satellites whose dimensions vary between a few meters and several thousands of kilometers across. Several thousand artificial satellites are revolving around the Earth.

■ **Shooting star** Brief streak of light produced by the entry of a small meteoric dust particle into the Earth's atmosphere.

■ **Sidereal period** Time taken by a star to return to the same position relative to the other stars.

■ **Solstice** Time of year when the Sun appears to stop its course of ascent or descent. The Sun then reaches the tropics, and the lengths of the nights are at their most extreme.

■ **Spectroscopy** Technique for breaking down light, thus yielding information about the chemical composition, the temperature and the physical state of the light source.

■ **Sublimation** Direct change from a solid to a gaseous state (without going through the liquid stage).

■ **Supernova** Rapid variation in the brightness of a star reaching the end of its evolution, which in a few days emits as much energy as several billion "normal" stars. The supernova phase is characteristic of massive stars at the end of their lives.

■ **Synodic period** Time taken by a star to return to the same position relative to the Earth.

■ **Tropical period** Time taken by a star to return to the same equinox, that is to say to the same point at which the plane of this star's equator intersects with the plane of the ecliptic.

■ **Variable (star)** Star whose brightness is not steady. There are different types of variable stars: periodicals, whose brightness varies regularly, and eruptive variables, whose brightness varies unpredictably.

■ **Vernal** The Sun passing the equator. The time when it occurs is known as the vernal equinox.

■ **Zodiac** Area of the sky in which the Sun, Moon and planets appear to move.

Appendices

Appendices

Appendices

Index

Appendices

Index

Photo acknowledgements

p. 4: NASA-HST; p. 8: NASA-MGS; p. 10: NASA-MGS; p. 12: NASA; p. 17: PARSEC; p. 18: Écomusée de La Roudoule; p. 19 top: Benvenuto-Minghelli; p. 19 bottom: Benvenuto; p. 20: NASA; p. 21: NASA; p. 22: NASA; p. 23: NASA-MGS; p. 24 left: ESA; p. 24 right: NASA; p. 25: NASA; p. 26: NASA-ESA; p. 28: ESO; p. 30: NASA; p. 32: NASA; p. 36 left: ESO; p. 36 right: NASA; p. 37: NASA; p. 38: NASA-JPL; p. 39: NASA-HST; p. 40: NASA-HST; p. 41: NASA; p. 42: NASA-ESA; p.43 left: ESA; p.43 right: NASA; p. 44: PARSEC; p. 46: NASA; p. 52 top left: ESO; p. 52 top center: NASA; p. 52 top right: ESO; p. 52 bottom left: OCA; p. 52 bottom right: ESO; p. 53 top: ESO; p. 53 bottom: PARSEC; p. 56: Benvenuto; p. 57: Benvenuto; p. 58: ESA; p. 59: Benvenuto; p. 60: ESA; p. 61 top left: ESO; p. 61 top right: NASA; p. 61 bottom right: ESA; p. 62: NASA-HST; p. 64: NASA; p. 66: NASA; p. 67 left: NASA; p. 67 right: NASA-HST; p. 68: PARSEC; p. 71: OCA; p. 73: PARSEC; p. 74: NASA-HST; p. 76: NASA-HST; p. 77: NASA-HST; p. 78: NASA; p. 79 top: NASA; p 79 bottom: ESA; p. 80: ESA-SOHO; p. 82: ESA; p. 84: PARSEC; p. 85: PARSEC; p. 89: OCA; p. 90: PARSEC; p. 91: ESA-SOHO; p. 92: ESA-SOHO; p. 93: ESA-SOHO; p. 94 top: ESA-SOHO; p. 94 bottom: ESA; p. 95: Viladrich; p. 96: ESA; p. 97: ESA; p. 98: ESO; p. 100: D. R.; p. 107: ESO; p. 109: PARSEC; p. 110: ESO; p. 113: ESO; p. 114: ESO-VLT; p. 115: ESO; p. 116: NASA-HST; p. 118: NASA-HST; p. 121: OCA; p. 124: Philip Parkins; p. 125: ESO; p. 126: PARSEC; p. 127: PARSEC; p. 128: NASA-HST; p. 129: Minghelli; p. 130: ESO; p. 131: ESO; p. 132: ESO; p. 133 top: NASA-HST-ESA; p. 133 bottom: ESO; p. 134: ESO; p. 136: ESO; p. 138: PARSEC; p. 142: ESO; p. 143: PARSEC; p. 145: DMSP-NASA; p. 146: NASA; p. 147: ESO; p. 148: NASA-HST; p. 149: NASA-HST; p. 150: NASA-HST; p. 151 top: ESO; p. 151 bottom: NASA; p. 152: NASA; p. 154: ESO; p. 156: Acuarez; p. 160: ESO; p. 163: PARSEC; p. 164: ESO; p. 166: ESO; p. 168: NASA; p. 169: NASA-JPL; p. 170: PARSEC; p. 172: NASA; p. 176: ESO; p. 178: OCA; p. 179: PARSEC; p. 181: PARSEC; p. 182: NOAA; p. 184: NASA-JPL; p. 185: ESA-ERS 1; p. 186: NASA; p. 187: ESA-CNES; p. 188: ESO; p. 190: PARSEC; p. 192: NASA-JPL; p. 193: NASA-JPL; p. 194: NASA-HST; p. 197: PARSEC; p. 198: PARSEC; p. 199 left: PARSEC; p. 199 right: AURA-KPNO; p. 200: Benvenuto; p. 202: Benvenuto; p. 203 top: PARSEC; p. 203 bottom: NASA; p. 205 top: NASA-JPL; p. 205 bottom left: NASA-HST; p. 205 bottom right: NASA-JPL; p. 206: NASA; p. 208: NASA-HST; p. 214: ESO; P. 215: ESO; p. 217: PARSEC; p. 218: ESO; p. 219: ESO; p. 220: NASA-HST; p. 221 top: ESA-HIPPARCOS; p. 221 bottom: ESO; p. 223 top: ESO; p. 223 bottom: ESA.